LEARNING GARDENS AND SUSTAINABILITY EDUCATION

Bringing Life to Schools and Schools to Life

Dilafruz R. Williams and Jonathan D. Brown

Routledge
Taylor & Francis Group

NEW YORK AND LONDON

D0609685

First published 2012
by Routledge
711 Third Avenue, New York, NY 10017

Simultaneously published in the UK
by Routledge
2 Park Square, Milton Park, Abingdon, Oxon OX14 4RN

Routledge is an imprint of the Taylor & Francis Group, an informa business

Library of Congress Cataloging in Publication Data
Williams, Dilafruz R., 1949-
  Learning gardens and sustainability education : bringing life to schools
  and schools to life / Dilafruz R. Williams and Jonathan D. Brown.
      p. cm.
  Includes bibliographical references and index.
  1. School gardens. 2. Environmental education. 3. Outdoor education.
  4. Experiential learning. I. Brown, Jonathan D. II. Title.
  SB55.W56 2011
  372.35′7--dc23   2011024463

ISBN13: 978-0-415-89981-9 (hbk)
ISBN13: 978-0-415-89982-6 (pbk)
ISBN13: 978-0-203-15681-0 (ebk)

Typeset in Bembo
by Cenveo publisher services

For Jesse
and
for Edward and Henry.

CONTENTS

Preface   *ix*
Acknowledgments   *xiii*

PART I
Learning Gardens, Living Soil, and Sustainability
Education   **1**

1   Introduction: Modern Imperatives for Life and Learning   3

2   Learning Gardens and Life's Lessons   15

3   Living Soil as Metaphorical Construct for Education   41

PART II
Learning Gardens Principles Linking Pedagogy
and Pedology   **55**

4   Cultivating a Sense of Place   57

5   Fostering Curiosity and Wonder   75

6   Discovering Rhythm and Scale   89

7   Valuing Biocultural Diversity   105

8   Embracing Practical Experience   120

9   Nurturing Interconnectedness   133

10 Awakening the Senses   146

PART III
Practice Comes Alive from the Ground up   **159**

11 Teacher, Principal, and Superintendent Perspectives   161

12 Moving Forward   195

Epilogue   *202*
Appendix: Selected Resources and Programs   *204*
Bibliography   *209*
Index   *218*

PREFACE

Sowing the seed,
my hand is one with the earth.

Wanting the seed to grow,
my mind is one with the light.

Hoeing the crop,
my hands are one with the rain,

Having cared for the plants,
my mind is one with the air.

Hungry and trusting,
my mind is one with the earth.

Eating the fruit,
my body is one with the earth.

Wendell Berry (1970a, pp. 57–58)

Learning gardens are an increasingly popular school-level undertaking, as they encourage healthy eating habits, require physical activity, foster biological diversity, improve soil and land stewardship, and nurture interconnectedness. In 2009, First Lady Michelle Obama invited school children to dig the grounds of the White House to plant vegetable gardens. By promoting gardens as one venue for student health, she has validated the school gardens movement. In a lifeless and concrete school building environment, learning gardens are a way to bring life to schools and schools to life. Furthermore, gardens provide an integrative context for lending meaning to education. This final quality is of significance considering the high school dropout rate despite many eras of educational reforms (e.g. *Back to the Basics; No Child Left Behind; Race to the Top*).

School gardens can serve as a common denominator for children to gain outdoor learning experiences. Given the recent interest in transforming education toward more sustainable practices where a wide range of efforts is weaving myriad opportunities that include harnessing wind and sun, purifying watersheds, recycling waste, and conserving energy, we posit that school grounds provide vibrant and meaningful learning sites to engage with the ideas and practices of sustainability. Making school grounds a healthy part of the educational environment requires a shift from paving with asphalt and/or manicuring with grass the vast land mass surrounding school buildings. It is in the actual process of conversion of these lots—from asphalt and grass to garden—that the life-giving qualities of soil can be appreciated and from which learning gardens grow. "In sight, in mind," soil surfaces as the frontier where nature, culture, and biology are intertwined, where humus teaches gratitude, where knowledge of de-composition becomes as significant as learning Composition.

Both authors have been involved in the creation of learning gardens and are part of a growing network of school garden educators. We have found school gardens as promising sites for nurturing lasting bonds of kinship with soil, creating what we term "soil kids." But, one might ask, what do gardens and living soil have to do with sustainability? The soil that constitutes the physical ground of learning gardens is home to plant, animal, and microbial life, and vibrant life itself, thus making it an exquisite entry point into teaching about relationships by breaking down ontological barriers that divide nature from culture, humans from non-humans, and food from soil. A dynamic food web, living soil exposes the fallacy of mechanistic understandings of life and calls upon us to re-member ourselves as part of the biotic community. As Thich Nhat Hahn says, within an interconnected world, "to be is to *inter-be*."

This book is an invitation to undertake a journey where we explore dominant and emergent metaphors, where life manifested in soil and learning gardens is at the center of the educational enterprise, and where a diverse sample of student writings and drawings roots the educational value of gardens in the grounds of learning for sustainability. As regenerative educators, we believe that this book can serve as a template for those who are concerned about the well-being both of nature and of children, and who see schools as living organisms rather than machines. Through articulating the attributes of living soil with corresponding pedagogical principles for sustainability education, we present learning gardens as a practical pathway for moving beyond dominant paradigms, bringing life to schools and, in doing so, bringing schools to life.

Organization of Chapters

Part I, "Learning Gardens, Living Soil, and Sustainability Education," situates learning gardens within a broader framework of sustainability that is now permeating educational institutions. Chapter 1, "Introduction: Modern Imperatives for Life and Learning," presents the dominant metaphors that guide modern educational reform efforts as background to several contemporary challenges that children face such as: lack of direct experience with nature; obesity, ill-health, and food insecurity; and ecological/environmental injustice. In order to move toward sustainability education, we argue that learning gardens at school sites can provide one tangible option to address these challenges. In Chapter 2, "Learning Gardens and Life's Lessons," we turn our attention to learning gardens, and draw upon historical trends, research, and current national examples of urban schools where gardens are flourishing. Our interest lies in asking the question: where is the learning in learning gardens? We explore the organization of learning gardens through a number of relevant national models that serve as a guide for design and development of curricula and pedagogy for successful integration of school gardens into education at various levels. In Chapter 3, "Living Soil as Metaphorical Construct for Education," we present living soil as a central metaphor and framework that supports transformation of modern educational systems through designing and developing learning gardens. We begin by recognizing the unfamiliarity of living soil and discuss in detail its role in sustaining and enriching life. This is of particular significance to gardens on school grounds which may be established in soils long buried under pavement or neglected. These soils need rejuvenation in order to support vibrant learning gardens and associated pedagogy. After exploring literal living soil, we elaborate living soil as a metaphorical construct highlighting seven specific attributes that can be used to guide learning gardens pedagogy and principles of sustainability education.

Part II, "Learning Gardens Principles Linking Pedagogy and Pedology," details each of the seven principles that link pedagogy and pedology (study of soil) via living soils of learning gardens: cultivating a sense of place, fostering curiosity and wonder, discovering rhythm and scale, valuing biocultural diversity, embracing practical experience, nurturing interconnectedness, and awakening the senses. In this section, Chapters 4 through to 10 interweave theoretical and conceptual discussion with abundant practical examples to illustrate possibilities. In order to address the de-contextualized nature of education in a globalized economy, Chapter 4, "Cultivating a Sense of Place," discusses the relevant construct of place.

Acknowledging the importance of place-based education, we problematize place seen as a panacea and propose an alternative dictum: "think ecologically, act locally" (Williams, 1995). In Chapter 5, "Fostering Curiosity and Wonder," we assert that direct experiences of nature in the learning gardens require a multifaceted awakening of children's latent urge to ask questions and to wonder. Chapter 6, "Discovering Rhythm and Scale," discusses ways in which learning gardens and living soil help to re-calibrate our sense of time and space. Chapter 7, "Valuing Biocultural Diversity," expands the discourse on diversity to include the intersections of ecology and culture. Given the demographic trends in urban schools toward increasing cultural diversity, we discuss ways in which learning gardens become inviting places where cultural traditions may be honored in relation to culinary and agrarian arts. Chapter 8, "Embracing Practical Experience," builds on the well-developed field of experiential education. We embrace the practical, given that soil must be experienced to be known. Chapter 9, "Nurturing Interconnectedness," explores patterns in nature and the concept of interdependence in nested systems. Chapter 10, "Awakening the Senses," invites us to engage our whole body in learning.

Part III, "Practice Comes Alive from the Ground Up," captures practitioner voices. In Chapter 11, we present "Teacher, Principal, and Superintendent Perspectives" relating how gardens are started, utilized as instructional tools, and supported by educators at different levels in their school systems. Chapter 12, "Moving Forward," is a celebration of life at the center of the educational enterprise. Leveraging the examples presented in Chapter 11, we revisit the seven principles that link pedagogy and pedology, reinforcing the connection between theory and practice and reiterating their interrelation to one another in representing the metaphoric construct of living soil in the learning garden.

We are both gardeners and have also founded and taught at several school gardens. We draw upon our experience as practitioners and also as researchers as we share examples and stories of how school gardens educate and what students learn in and through gardens. Utilizing an interdisciplinary conceptual framework, our intent is to ground learning gardens in life's marvels and mysteries over which no single discipline can have explanatory monopoly. When guided by a theoretical foundation, school gardens have strong roots in academic learning and are integrated into school life; they are not viewed as "add on" or as "after-school" ventures and activities. In joining hundreds of educators who have embraced learning gardens as one pathway toward sustainability, we are encouraged by their life-affirming practices.

While written in large part for an academic audience, this text seeks to bridge theory and practice, as learning gardens come alive in the voices of students and educators and in the diversity of examples across schools and particularly across large urban school systems in dire need of providing children and youth with experiences of nature.

ACKNOWLEDGMENTS

Soils and gardens come to life through innumerable relationships. Likewise a book arises from countless conversations, inspirations, experiences, and partnerships, with people young and old, across cultures and institutional affiliations, in school rooms and on school grounds.

From an ecological perspective, ideas emerge through the life flow of an interconnected complex web of relations. This book has been several years in the making. Where an idea is seeded and how it germinates is not always easy to pin-point—it is in the practical engagement with ideas and with people that a book such as this comes to life. We are humbled and grateful to be part of such engagement with people and place, soil and gardens, both in Portland and across North America in urban school districts.

The stories captured in this book honor the voices of children and youth, garden educators and classroom teachers, school and district administrators and individuals in governmental and non-profit organizations whose commitment to and love for this work are strongly rooted and grounded in nature and gardens. We cherish the chance to tell these stories.

We are honored to bear witness to a movement that takes the risk of going against the grain of modernity and champions school gardens for learning. While the influence of all the individuals upon this book cannot be fully captured, we are particularly grateful to the following friends and colleagues.

To teachers Jennifer Andersen, Deziré Clarke, Chad Honl, Angela LeVan, Peter Leki, Robin May, Abby Roth, Rebecca Wagner, Robert Wright, and countless others, who understand the value of school gardens to learning as they integrate them with curriculum.

To hundreds of children and youth of all ages who delight in the wonder of sprouting seeds, who resist and later revel in fresh garden produce, who plant dreams in the ground, who remind us to play, and who, like wriggling worms, breathe life into the living soil of school gardens.

To school and district leaders Carlos Garcia, Tim Lauer, Karl Logan, Teri Sing and other principals and superintendents, who take risks championing the beliefs that good education is experiential, engaging, and nourishing for the soul. And to Sarah Taylor who, as founder and principal of Sunnyside Environmental School, embodies the sentiment that permeates the pages of this text through her work in support of students and teachers.

To founders, leaders, staff, and school garden supporters of organizations including Denver Urban Gardens and Slow Food Denver; Urban Harvest in Houston; the Learning Gardens Laboratory and JEAN's Farm in Portland, Oregon; Green Schoolyard Alliance and Urban Sprouts in San Francisco; Intergenerational Landed Learning at the Farm, Vancouver BC, who bolster school garden initiatives and provide outside knowledge, skills, and assistance.

To garden educators including Judy Bluehorse, Gregory Dardis, Kara Gilbert, Julia Hamlin, Madelyn Mickleberry-Morris, Stephanie Rooney, Marcia Thomas, and Valerie Thompson, who support communities, schools, teachers, and students in the rich living milieu of learning gardens.

To Pramod Parajuli, a visionary leader, who brought together a community of scholars, educators, and students under the umbrella of the Leadership in Ecology, Culture, and Learning (LECL) program at Portland State University; who is dedicated to promoting edible and multicultural gardens, to regenerate urban foodsheds and learningsheds. This work represents the delicious harvest of the seeds planted during the LECL era.

To intellectual ancestors that inspire our thinking and action; most notably to Wendell Berry, who encourages us to teach our children to grow food and connect with land wherever we live; and to Gandhi, who reminds us that to forget to dig the earth is to forget ourselves.

To the Spencer Foundation and Portland State University Faculty Development program for grants awarded to Williams for research and writing.

To Larry Beutler, editor of *Clearing Magazine* and Larry Frolich, editor of *Journal of Sustainability Education,* for permission to use our previously published essays.

To professional contacts who shepherded this work toward publication and especially to Naomi Silverman of Routledge for truly understanding the worth of this work.

To the anonymous peer reviewers whose comments strengthened the work, and to the editing and production staff, particularly Kevin Henderson at Routledge, who have materialized our visions upon the printed page.

To family, teachers, students, and friends, for their grace, flexibility, and willingness to share, and whose contributions have made such a celebration of life and learning possible. To Wayne, for the field behind the plow; to Ted, for introducing *Mindwalk*; to Adam, for radical homemaking; and to Molly, for care, encouragement, and companionship in the garden. To those closest to us: to James, for recognizing the worth of good soil and compost as "real gold," and for endless patience, counsel, and guidance in bringing this book to life; to Mary Anne, for delight in the educational value of learning gardens and for editing support; to Gulnar, for technical assistance with models; and to Janice and Richard, for being willing to experiment with worms.

Finally, we appreciate each other for robust dialogue in cultivating soil and soul in the garden manifested in this book.

PART I

Learning Gardens, Living Soil, and Sustainability Education

1

INTRODUCTION

Modern Imperatives for Life and Learning

Sustainability education is now an emergent field of possibilities; gathering hope toward a "climate of change" in education. Green school and university initiatives include sustainability offices and centers to address policy and practical matters related to: building construction, transportation, energy usage, nutrition and health, recycling, schoolyard habitat, waste reduction, water harvesting, among others. The United Nations has framed sustainability and sustainable development as global goals for education (UNESCO, 2010). Sustainability indices are being developed and institutions are "ranked" based on their adherence to sustainability criteria addressing economic, environmental, and social issues.

However, amidst a generally positive atmosphere encompassing a wide range of vibrant multidisciplinary sustainability interests, we find that the fundamental assumptions guiding curriculum and pedagogy are left unaddressed. There remains a lingering tendency to continue to enshrine uniquely modern ways of thinking. Partly this is achieved through repetition of dominant metaphors, as well as through carrying forward un- or under-examined cultural, epistemological, and ontological assumptions that encourage piecemeal, rational, and detached "objective" views of the world (Bowers, 1997; Esteva & Prakash, 1998; Kumar, 2002; Sterling, 2001). It is inevitable that even the concept of "sustainability" would succumb to the rules of opportunistic engagement (Sauvé, Berryman & Brunelle, 2007).

As an umbrella term, "sustainability" is now touted despite many philosophical contradictions as seen in the following examples:

- advertising and marketing "sustainable" eco-harvested forest logs from Belize for use in constructing houses in clear-cut Appalachian mountains;

- securing LEED "sustainability" certification for monumental skyscrapers in New York;
- building large-scale "sustainable" wind farms off the pristine coast of Cape Cod;
- "sustainably" growing organic bananas produced as monocrops on large farms in Hawaii owned by absentee corporations and shipped to monopolies of food-chains thousands of miles away on 48 mainland states;
- securing mega-factories for the production of "sustainable" solar panels in Shanghai causing enormous environmental waste;
- sending tons of biomass daily from chemically manicured yards in gated communities to mega-composting "not-in-my-back-yard" dumps.

The list goes on. Basking in the glory of such "sustainability" endeavors, there is little recognition of their undergirding consumptive capitalist and dominant paradigms. The Eden of sustainability is often informed by the same modern mindset that has created our educational systems often tied closely to the global economy. Thus, the emerging discourse on sustainability education is in many ways caught in a modern web of theoretical, ontological, and epistemological assumptions that are incongruent with sustainability.

Modern Education Reforms

Each year, millions of children in the United States spend about ten months in a school. No matter what the age of the child, much of this time is expended at desks learning the "basics," that is, the three Rs of reading, 'riting, and 'rithmetic, in the classrooms within the confines of the four concrete walls of the school building. The pressure of global competition continues to escalate further and, as a result, many schools are even doing away with recess and play time. Time spent outside is considered non-essential.

For over half a century, the fear of America losing its dominance as a super power in the world has led to an overcrowded and packed curriculum in hopes of addressing global competition. There have been waves of education reform couched within the language of *risk* and *crisis*. In the 1980s, President Reagan declared that America was "a nation at risk"; at that time the manifesto was about the drop in test scores and academic underachievement across national and international scales, with Japan as our main economic competitor. We have not recovered from the metaphor of American schools being *at risk*. As an evocative term, it conjures fear, danger, concern, and often pity and sympathy for those segments of the student population that the system has been unable to serve.

In order to deal with the crisis of American students not being competitive among industrialized nations, there have been a number of recommendations proposed, among which are raising expectations, increasing time for instruction, and using market-driven ideas to inform performance. *A Nation at Risk* was precursor to *No Child Left Behind* (NCLB). For a decade now, NCLB has rightly pointed out the race and class differences in academic achievement in our schools; NCLB forced schools to laser focus on closing the achievement gap in Reading and Math. However, the mandates of testing have come with the threats of punishment. So what do public schools do? Test children over and over again until they get the one correct answer on a multiple choice test? Filling in bubbles of predetermined answers rather than bubbling with the joy of learning is the consequence for children. Top-down reform and standardized tests for measuring achievement have become the stick that drives educational policy and practice of reform.

Since 2009, the reform used at schools is *Race to the Top*. Driving this agenda is the worry that America is losing its clout in global competition, this time to China and India, and that we are dropping in stature in our academic performance with Finland topping the list of performance and other countries racing faster than we are. Stanford scholar Larry Cuban (1990) explains it best when he writes:

> The return of school reforms (again, again, and again) suggests that the reforms have failed to remove the problems they were intended to solve. Analysts ask: Are we attacking the right problem? Have the policies we adopted fit the problem? ... Were the solutions that were designed to correct the problems, mismatched? Right problems, wrong solutions? Or vice versa? Are we dealing with the problem or the *politics of the problem?*
> *(pp. 5–6, emphasis added)*

More important, we believe, is to ask ourselves, what metaphors are being used as a basis to shape and reform policies and drive educational practices? Often, marketplace analogies and business models argue for greater efficiency. Yet, with the 2009 Wall Street and global market collapse, there is not much trust in these institutions. Moreover, market and business models are themselves predicated upon lifeless *mechanistic* metaphors.

Guiding Metaphors of Modern Education

Discussing the power of metaphorical thinking, Maxine Greene and Morwenna Griffiths (2003) in their essay, "Feminism, Philosophy, and Education: Imagining Public Spaces," explain:

> We need to rethink, to think differently: to use our imaginations again ... metaphorical language [is] a way of rethinking and questioning orthodox thinking. A metaphor is what it does. A metaphor, because of the way it brings together things that are unlike, reorients consciousness.
> *(p. 85)*

At present our educational system is guided by lifeless mechanistic metaphors, manifested in a conceptualization of schools as complex machines. As Greene and Griffiths (2003) contend, a metaphor is what it does; thus, guided by mechanistic metaphors, schools progressively mimic machines in form and function. We present the following seven areas of concerns to highlight the troublesome trends that emerge from a mechanistic orientation.

1   *De-contextualization of learning.* Education is too often framed with little regard to the physical places in which schools actually exist. Though schools exist in physical communities, filled with social and biotic actors and relationships, education follows a de-contextualized script (curriculum) without reference to the local context: this cripples the ability of the educated to "see the forest for the trees." The intricate web of relations to which the tree is bound is obscured by the "silo-ing" (Orr, 1992) tendency of the academy. Within this view, knowledge retains little relationship to the social and ecological context from which it arises and in which it must be ultimately applied. Responsible application of knowledge—that is, considering potential ecological consequences of actions—is overlooked in such a de-contextualized transmission of "neutral" knowledge (Capra, 1996; Orr, 1992).

2   *Loss of curiosity and wonder.* Why do children enter school as a question mark and leave as a period? It is unnecessary that children's latent urge to ask questions be stifled by standardized tests and the "get it right" syndrome of modern education. Within the structure of NCLB high-stakes testing, time for curious exploration and unbound wonder is curtailed to ensure children and youth are able to master multiple choice tests in which there can be only one correct answer among a set number of options. The dichotomies characteristic of the modernistic Cartesian paradigm are carried forward in sets of binaries germane to education such as right–wrong, teacher–student, and teaching–learning. These separations promote oppositional arrangements that privilege the teacher as all-knowing and position the student as a passive receiver of transmitted knowledge. Such simplistic scenarios are rarely repeated in real life and do not positively contribute to joyous learning.

3   *Acceptance of mechanical and industrial scale.* Modern education is set in lock-step to a rigid clock: hours of learning are regulated by the lifeless ringing of bells, and years of mental, physical, and emotional development are charted in stages of linear development. The life rhythms of natural cycles are ignored or overshadowed by the beat of an industrial march ever forward. But forward toward what? Additionally, for too long the dominant trend in modern Western culture has been to think "big" (Berry, 1970b). This proclivity toward grandiosity continues to influence education and the emerging sustainability paradigm resulting in the discounting of multitude of diverse human and biotic communities specific to particularities of place.

Large-scale, top-down actions are taken for granted. Thinking big often turns out to be not thinking wisely.

4   *Homogenization of curriculum and learning.* Homogenization weakens ecosystems and social systems, whereas diversity strengthens them through building complex networks of interdependence. At present, the homogenization of curriculum emphasizes the industrial quality of schools in which the critical importance of context is erased. The production and transmission of knowledge is divided spatially, socially, and temporally from society and removed from the local human and biotic communities in which schools physically exist (Smith & Gruenewald, 2008). Racing to the top on the road to progress, children's latent creativity, curiosity, and wonder are paved through the explicit standardization and uniformity of curriculum and learning methodologies with a view toward ever more efficient use of human, informational, and intellectual "resources."

5   *Privileging of abstract ideas.* Modern educational systems divorce knowledge from lived experience and affective dimensions of life. All learning is expected to take place between the ears as the body sits still for seven to eight hours per day. The claim made by many students at all levels that school is "irrelevant" to life rings true considering that information and knowledge are continually abstracted from application. Modern educational systems divide knowledge into smaller and smaller specializations, such that experts are produced knowing little beyond their realm of specialization and having little experience in application. A privileging of abstract knowledge separates head from hands, mind from matter, and ideas from experience (Sipos, Battisti & Grimm, 2008; Sterling, 2001). Within modernistic educational systems, there is a clear delineation of status ascribed to knowledge. High-status knowledge is abstract, theoretical, scientific, and de-contextualized from the physical world. Low-status knowledge is associated with manual, craft, or trade knowledge, and has typically been limited to high school vocational training and community colleges (Bowers, 2000). An inequitable pattern of funding provided for techno-scientific research while humanities budgets are reduced demonstrates a division of values in monetary terms. One of the dangers of privileging high-status abstract knowledge at the expense of practical place-based knowledge relates to devaluing forms of cultural capital encoded in oral traditions and marginalizing face-to-face, recursive, iterative, experiential, temporal, spontaneous, and long-term teaching and learning relationships embedded in local cultures and ecologies (Cajete, 2001; Smith & Williams, 1999).

6   *Perpetuation of individualism and autonomy.* Subject matter in schools is abstracted from relation within a broader picture of whole-systems knowledge. A number of European Enlightenment ideas such as Descartes' declaration of independence, "*cogito, ergo sum,*" or "I think, therefore I am," Charles Darwin's theory of evolution, which cast the individual as the

basic ecological unit and the natural world as a battleground for scarce resources, and Adam Smith's notion of the "invisible hand" guiding capitalist economics through rational self-interest (Esteva & Prakash, 1998) deepen an ontological division of mind from matter and culture from nature, and emphasize an individualist outlook on existence (Kumar, 2002). In terms of educational practice, an honoring of the autonomous individual at the expense of community interconnectedness encourages a competitive approach to achievement (e.g. *Race to the Top*) even when couched in terms of collective movement (e.g. *No Child Left Behind*).

7   *Stimulation of only certain senses.* Modern educational systems divorce knowledge from lived experience and affective dimensions of life. An overstimulation of the eyes and ears without exercise of the other senses diminishes the bodily value of learning. It is not uncommon for children and youth to spend upwards of eight hours per day engaged with a computer or television screen, while time spent exercising other senses is minimized. Our senses of smell, taste, and touch are closely associated with deep learning and memory: we are more than just minds, eyes, and ears.

Taken together, these seven aspects of the modernist orientation—de-contextualization of learning, loss of curiosity and wonder, acceptance of mechanical and industrial scale, homogenization of curriculum and learning, privileging of abstract ideas, perpetuation of individualism and autonomy, and stimulation of only certain senses—are incongruent with living systems and sustainability. Founded upon mechanistic metaphors, educational reforms imagine schools as no more than complex machines and overlook the value of life itself. What is sorely missing in the discourse of educational reform is an understanding of the power of guiding metaphors and acknowledgment of the link between landscape and mindscape. In bringing sustainability to education we must address the physical environment *and* transform dominant mental models that underwrite connection with the land. Contextualized understandings and holistic relationships among tangible living entities are hallmarks of sustainability (Capra, 1996), thus a metaphoric framework that is more ecologically grounded is needed. School grounds and schoolyards are prime milieu that serve as the basis for ecological alternatives that contrast dominant mechanistic metaphors. Not only can learning gardens enhance mastery over literacy, numeracy, as well as life skills, but for us, soil and learning gardens also serve as animate options for reorienting the metaphoric imagination guiding modern education.

Modern Challenges, Modern Imperatives

The convergence of the following strands of public concern related to personal, social, and environmental sustainability surfaces school learning gardens as one dynamic example of a whole-system solution:

Obesity, ill-health, and food insecurity: There is an elevated interest in teaching students how to grow and be connected to local sources of food. This is particularly so given the following trends: childhood obesity rates at an all-time high; pandemics such as the swine flu and waves of salmonella outbreaks causing public frenzy; and type 2 diabetes on the rise especially affecting non-white youth populations (Fagot, Burrows, & Williamson, 1999; Hedley, Ogden, Johnson, Carroll, Curtin, & Flegal, 2004). For example, in 2007, more than 16% of children and adolescents aged 6–19 years were overweight, a 45% increase from 1994. Obesity is a major contributor to diabetes, heart disease, and other preventable diseases. Part of this has to do with reduction in healthy food choices. Less than 20% of children are eating recommended quantities of fruits and vegetables: 20% eat no vegetables, and 35% eat no fruit. It is predicted by the United States surgeon general that children today may be the first generation to have a shorter life span than their parents (Winne, 2008). Since small farms have been steadily disappearing and there is an increased globalization of food systems, children and a generation of adults do not know where their food comes from, how it is grown, or how food gets on their plates. Food is seen more as a commodity than as a life-giving entity. In response to this disconnect and as a vivid example of rising local food consciousness, there is a growth of farmers' markets across the country. In addition, food gardens have emerged as a critical area for localized action in response to global environmental degradation as they are identified as practical means to reduce food transport miles, conserve water, reduce pesticide use, etc. Via learning gardens, then, there is a perceived opportunity to teach children about sustainable food systems in response to a growing disconnect from the sources of our food. These gardens are seen as a means of reviving food knowledge so that children can make appropriate decisions about what they eat as this impacts their health. Participation in school gardens has been linked to increased familiarity with and consumption of fruits and vegetables (Ozer, 2007). Thus, in order to connect children and youth with their sources of food and related health matters, school grounds are considered prime places for learning.

Nature Deficit Disorder and No Child Left Inside Initiative: A reduction in children's access to natural spaces and engagement with more-than-human phenomena contributes to what Richard Louv (2008) has termed "nature deficit disorder." A number of scholars observe that, today, many children have limited contact with natural systems such as forests at the edge of cities or even abandoned un-paved lots (Louv, 2008; Chawla, 2006; Orr, 1994). Louv's (2008) warnings about saving our children from nature deficit disorder has spurred the formation of a large national Children and Nature Network with a sense of urgency to offer children outdoor spaces to play and experience nature. It is not uncommon for children to spend less than 5% of their time outside, while many routinely spend upwards of eight hours per day engaged with a computer or television screen. Children in urban areas in particular have limited contact with nature.

Lacking sufficient relevant exposure to and experience with natural systems, how will today's students be able to develop the requisite ecological literacy likely needed to negotiate future ecological challenges? As an antidote to the *No Child Left Behind Act*, which is seen as narrowly defining curriculum and restricting children, a *No Child Left Inside* coalition has emerged. Informal contact with natural systems can support children's emotional growth, provide solace during emotional years of adolescent development, as well as profound allegorical models for expressing personal, social, or academic transformation. Complex changing cultural norms and patterns of settlement make such formative informal contact ever more difficult to experience (Kunstler, 1993; Louv, 2008). School gardens can provide the grounds for daily formative contact with living things for students otherwise alienated from the natural world.

Concern about ecological/environmental injustice: Recognition that ecological degradation—ranging from loss of fertile topsoil by erosion to poisoning of lakes, rivers, oceans, and fisheries by toxic substances—is intimately interconnected with social injustice and is now becoming harder to ignore. In too many cases, the question is no longer "if" but "when" rivers will run dry, fields will go fallow, and oceans will rise. Accompanying vast disruption of natural systems is the dissolution of the social safety net in the era of neo-liberal austerity budgets. In the shadow of the Great Recession, an increasing number of Americans find themselves falling through holes in the metaphorical safety net, joining the ranks of the unemployed or underemployed, while wealth continues to be ever more concentrated among a small number of elites.

While *New York Times* columnist and best-selling writer Thomas Freidman (2005) has declared that, in the context of economic globalization and trade liberalization, "the world is flat," casual observation of the daily headlines suggests to the contrary of his central hypotheses, the world seems to be becoming *more* hierarchical and prone to war, and *less* transparent, democratic, and prosperous (Varadarajan, 2005). Eco-feminist Vandana Shiva (2005) has remarked that the corporate flattening process Friedman celebrates as a force for global unity overlooks an unprecedented division of people along lines of economic class, geography, race, gender, religion, etc., and destruction of the web of life. Extensive studies have documented a consistent and growing pattern in the locating of environmental hazards, such as toxic waste dumps, in low-income neighborhoods where communities of color reside (Chiro, 1996; Jones, 2007, 2008; McLaren & Houston, 2004). People who exist in direct relation to threatened places are themselves becoming endangered. The erosion of topsoil, for example, has been accompanied by an erosion of vibrant and diverse farming communities (Berry, 1970b; Montgomery, 2007).

Heightened attention to climate change among the general public has opened opportunities to explore sustainable options in every sphere of life. It has generated interest in conserving farm land, revitalizing deforested areas to protect soil and waters, and curtailing loss of biodiversity. Much of the concern centers on ways that reclaiming the local and connecting to the source of food, water, and soil are critical to the health of people and planet, alike. Learning gardens emerge as one example of a whole-system solution that addresses these concerns and can be implemented directly on the school grounds.

Learning Life's Lessons

A disconnection of education from life undermines the relevance of education to life. Lessons related to the cycles of life and death, growth and decay, and renewal against obstacles cannot be taught through video games. Moreover, life's lessons are absent in the consumer culture of the mall and difficult to discern on *Facebook*. Despite decades of educational reform efforts, schooling today remains irrelevant to many students, as evidenced by the growing high school dropout rate. Yet, as Dewey (1916) reminded us, education is not only *about* life, truly meaningful education *is* life. We challenge the mechanistic metaphors that guide modern schools and look toward an alternative ecological paradigm that is more life affirming. As islands of biological and cultural activity, school gardens bring a welcome breath of fresh air to the learning environment. Planting together (as seen in Figure 1.1), children's hands filled with rich organic soil (Figure 1.2) can literally bring life to schools.

FIGURE 1.1 Harvesting bounties of learning gardens

FIGURE 1.2 Tactile literacy in action: a scoop of soil knowledge

These are precisely the types of life lessons—that is, lessons about life—that are needed today. Within the context of an increasingly urbanized population and school district budget cuts that make visits to nature centers and outdoor schools difficult, school grounds themselves can become sites for taking part in the transformative processes of nature, and bringing children into membership with soil and with the biotic communities in which schools are situated. For children and youth, where the common denominator of experience is school, there can be no better place to commune with life than below the feet on school grounds.

Filled with blossoming flowers and shrubs, ripening plants, and decomposing compost heaps, gardens located on school grounds provide immediate and consistent exposure to natural systems. And as examples of sustainable interface between natural ecological and human cultural systems, learning gardens demonstrate that a right relationship is possible between nature and culture. Neither an exotic wild area nor an entirely human-built landscape, gardens represent a diverse interstitial place in between biosphere and "culturesphere" (Parajuli, 2001). Seen in this way, school gardens offer not only an antidote to "nature deficit disorder," but also substantial restoration of relocating the human place in the natural world one school yard at a time.

The reductionist, mechanistic, industrial framework that has guided modern Western capitalist social institutions is now straining to negotiate the innumerable social and ecological challenges that characterize the early years of the twenty-first century. None of the dilemmas of our time can be understood in isolation from one another: mechanistic solutions promoted by the dominant modern paradigm have more often than not exacerbated the very problems they purport to solve. An ecological perspective is needed to contextualize seemingly disparate crises as interconnected parts of a whole. The emerging discourse on sustainability endeavors to develop a language to express these relationships.

We believe in reorienting the guiding metaphors of education vis-à-vis school gardens to broaden and deepen the understanding of the nature of education with a primary focus on the relationship between learning and life. Forging beyond a critique of modern education, we place *life* at the center of the enterprise of schooling. To this end, we use the metaphor of *living soil* as a foundation for designing and using learning gardens to also engage questions such as: What is education for? What is learning? What is an educated person? What is the connection between school and community? How can we bring life to schools?

Soil and Life

Specifically, *living soil* serves as a potent metaphor which contrasts that of the machine. Rather than racing headlong toward "the top," and forgetting to forge any commitment to place, we instead propose a re-grounding of learning in place through cultivation of gardens on school grounds. Thus the final frontier of pedagogy lies not in the distant cosmos but underneath our feet in the living soil of school learning gardens.

In an increasingly urbanized society, soil is often "out of sight, out of mind," paved under asphalt and concrete. Children, particularly urban children, grow up surrounded by gray roads and concrete buildings rather than engaged with the richness of living soil. They are more likely to recognize the aroma of tar, not soil. Human cultures have had historical, spiritual, and sensual relationships with soils (Kumar, 2002; Shiva, 2008). Thus, paving over soil alters the human experience and psyche in deep ways. Unlocking the mysteries of soil and restoring human–soil relationships help us to unlock the secrets of life. As a fundamental component of all life, soil is likely a genuine predictable indicator of sustainability.

Terms such as *earth* and *ground*—while some of the oldest in human language—are etymologically related to *soil*. In addition, words such as *humus*, *humility*, and *humanity*, are related linguistically. Soil is intimately connected with human culture and history (Hyams, 1976; Montgomery, 2007). Through soil, we learn about the sacredness of life, as taught by such diverse texts as the Gita or the Bible. As a living entity, soil invites us into kinship and serves as more than a mere growing medium. Soils are a web of relationships, heaving with countless digestive systems, nervous systems, and skeletal systems. When we are able to get to know a particular soil, as in a garden of long tenure or a home bioregion, we often discover that soils can even have personalities. It is necessary that children regain a tangible understanding of soil; school learning gardens create dynamic opportunities for students to engage with and be engaged by living soil. Here, we seek to develop attributes of living soil into a metaphorical framework that can guide education toward an ecological paradigm for sustainability, from which emerge seven specific pedagogical principles: cultivating a sense of place, fostering curiosity and wonder, discovering rhythm and scale, valuing biocultural diversity, embracing practical experience, nurturing interconnectedness, and awakening the senses.

What is Education for?

At the heart of this discussion, is the fundamental question: "What is education for?" We believe that schools can enrich life: the life of children, the life of families, and the life of human and biotic communities. In this, soil plays a significant role, an assertion which may cause trepidation among educators who, like their students, may be unfamiliar with soil. The beauty of soil as a metaphorical guide is that it opens up a universe of questions to which there may be no finite answers, and shifts the emphasis of education from teaching to learning. For guidance, we turn to the poet Rilke (1934), who reminds us: "Be patient toward all that is unresolved in your heart and try to love the *questions* themselves ... And the point is, to live everything. *Live* the questions now. Perhaps you will then gradually, without noticing it, live along some distant day into the answer" (pp. 34–35, italics added).

In order to develop an understanding of the fundamental issues of education, we enter into a dialogue with thinkers who have challenged the status quo and have shown the ways less travelled. The rich tradition of educational thought that guides our work is derived from Wendell Berry, C.A. Bowers, Fritjof Capra, John Dewey, Gustavo Esteva, Satish Kumar, David Orr, Madhu Prakash, and Vandana Shiva, among others. Education can challenge the very core of what it means to be *human* in the twenty-first century. Instead of preparing kids for future employment and a career driven by technological goals, schools can, instead, focus on the present health of our children—emotional, physical, social, spiritual, and mental. Learning gardens serve as one, though not the only one, means for such an education.

2

LEARNING GARDENS AND LIFE'S LESSONS

In this chapter we establish learning gardens as legitimate academic venues through surveying national model programs and exploring theoretical frameworks that help locate and explain the learning that occurs in the learning gardens. While learning gardens have emerged as a popular school-level activity, lacking a thorough theoretical framework, they risk fading away as yet another fanciful educational trend.

While we ground learning gardens as rigorous academic venues that enhance learning, we emphasize ways in which they shift the underlying metaphorical orientation of modern schools. Students, teachers, and communities use learning gardens as pathways toward sustainability education at many levels, moving toward partnerships that link life with learning, schools with neighborhoods, neighborhoods with bioregions, and nature with culture. Below, consider students' quotes about their experiences in school learning gardens.

> It is important to listen to nature because nature is so beautiful to look at! Why shouldn't it be just as beautiful to listen to?
>
> *(3rd grade student)*

> Today we tasted rainbow chard. The dark green leaves tasted kind of bland but the yellow or red stems tasted really good—kind of sweet in a really cool way. They also tasted fresh. We also tasted a kind of yellow flower called *nasturtium*. It had a dark yellow center and the yellow part was sweet and the red part was spicy. I also watched a spider rebuild its web.
>
> *(4th grade student)*

> The Learning Gardens is a time to be in your own little world. Letting your imagination go wild. Planting dreams in the ground and see them grow. If I can do this ... take care of a plant, then I can see that I can take care of anything. I can take care of myself and help myself and others.
>
> *(6th grade student)*

"When we try to pick out anything by itself, we find it hitched to everything else in the universe," wrote John Muir (1911). As seen in the student voices above, Muir's observations are manifested daily by students sensitized in ways that only nature teaches: these third, fourth, and sixth graders are learning to observe, to taste, to smell, to listen, to touch, and to care. They are able to discern and differentiate, imagine and connect, all the while learning about life directly from soil and nature in the gardens that they can step into right outside the doors of their school. One student joyously reminds us that nature is more than just a visual phenomenon to be seen. Beyond her suggestion that we also listen to nature is a nuanced understanding of the depth of relationship possible among humans and the environment. A second student deliciously describes tasting a new vegetable—rainbow chard—using fine details to distinguish among various parts of a new plant. Increased exposure to novel foods can improve children's eating habits; moreover, this quote demonstrates cultivation of exquisite observational skills in tandem with new taste buds. Another student makes a beautiful metaphorical leap, describing "planting dreams in the ground and [seeing] them grow." This student links the skills that he is learning in the gardens with the life skills needed for care of self and community. These brief quotes from students in learning gardens show multiple levels of learning and meaning making beyond memorization of facts and figures. For example, in Figure 2.1, students draw upon their study of geometry as they lay out plants in a hexagonal planting pattern. In contact with the living world in learning gardens, these students are broadening their understanding of life's mysteries even as they gain academic skills.

This is in stark contrast to much of education vis-à-vis schooling where teaching and learning are enclosed within the four walls of the classroom supported by wood and bricks, steel and glass, separating the exterior of the building from the interior. The physical structure of the building is interrelated to an overall educational paradigm, transmitting a hidden curriculum that often ignores life and is disconnected from the surrounding community.

To counter the manifest and literal lifelessness of schools and classrooms, since the early 1990s an emerging national movement in the United States has been focusing on shaping the school grounds hitherto covered with asphalt or manicured lawns into green schoolyard habitats and school gardens, a prototype of which is seen in Figure 2.2.

This resurgence of interest over the last 20 years has resulted in the establishment of thousands of school gardens across the country. Given the

FIGURE 2.1 Digging and planting

FIGURE 2.2 Students planting in a newly transformed garden: a year prior, this site was an abandoned school lot

interdisciplinary nature of gardens and garden education, this is to be expected. Simultaneously, garden-based learning is being supported by curricular efforts to align standards particularly in science, mathematics, language arts, nutrition, geography, literature, and health science, along with skills acquisition. Support for the design of gardens spans rural, urban, and suburban school districts. Subject area standards are also aligned by states to meet the newly designed curricula to ensure consistency across districts for garden-based learning (Barrs, Lees, & Philippe, 2002; Blair, 2009; Ozer, 2007).

Learning gardens on school grounds provide poetic and critical texts for nurturing students' connection with the more-than-human world. In the following extract, Molly, a third grade student in an inner-city school, describes how her class converts a trashed empty lot on school grounds into a garden. While reading Paul Fleischman's (1999) *Seed Folks*—a small book that presents big lessons about community transformation through the collective conversion of an abandoned lot into a garden—the class learned about soil and food, about clearing the vacant school lot, about building garden beds, and about planting and harvesting. Along with her classmates, Molly is directly engaged in bringing forth the life-giving qualities of her school gardens as the class builds the gardens and as transformation becomes the key to understanding life as it emerges.

> [Our class garden] was an empty, trashed lot. Nothing went on there but some stray cats meowing and leaf debris falling to the bleak, grey, concrete ground. This story is about transformation—from a bleak locked up place to a wide-open group of garden boxes filled with beautiful baby plants.
>
> Transformation. A lot of this happened with a "simple" garden. We picked up all the leafy debris on the ground and put it into compost: good soil for the earth! The lot was starting to transform…
>
> Trash! Dirty old litter. Disgusted hands picking up gum wrappers and old, de-toothed combs. We finally finished. The transformation was complete.
>
> However, we needed to bring out the lot's good side, not just destroy its bad side. That requires another transformation—seed into plant. The tiny black onion seed turning into an onion? All of us shook our heads in doubt. However, after planting the seeds in egg cartons filled with fertilizer, a fourth transformation began to happen—and the most wonderful and magical of all. A sprout appeared. A sign of life out of a tiny, dead-looking thing. We all shook our heads again, this time in amazement. The same happened with many other types of plants. Beets, radishes, carrots and more went into egg cartons. Each sprouted. Each transformed.
>
> Our class had come to transplanting to the transformed lot. Little seedlings went into dirt-filled, garden boxes. The lot was actually becoming fun! …

Our botanist taught us everything about transplanting, composting, mulching for the winter. She was a big part of our garden.

Then we decided to pull up some radishes to make a salad. We forgot about it, so they shriveled up. That was a big blow.

But we had made a complete transformation happen from a dirty lot … and brought out the lot's good side.

(3rd grade student Molly)

Following the format of Fleischman's *Seed Folks*, which unfurls the plot through the eyes of individual characters, the students wrote a narrative of their own community transformation by linking individual stories into a small book they called *Seed Kids*. In the extract above, Molly tells the story through her perspective, explaining how the class brought out "the lot's good side" through transforming seed into plant. Her compelling story encourages us to remember our capacity to positively contribute to life. She and her classmates turn leafy debris into compost: "good soil for the earth!" Moreover, she and her classmates turn their school environment into a living laboratory that supports their learning and development. For children and youth, where the common denominator of experience is school, it is the school grounds themselves that can be the context for engaging in such deep transformation, meaning making, and learning.

Guiding Sustainability Pedagogy: A Partnership Model

A key component of learning gardens is an abundance of partnerships between and among diverse social, cultural, ecological, and generational groups. In Figure 2.3, we present a Partnership Model of Sustainability developed by Pramod Parajuli (2002).

This model informs our learning gardens design, curriculum, and pedagogy; it also guides the articulation of learning gardens as whole-systems design for sustainability education. Parajuli explains that solutions to the complex, multi-faceted problems in the world—particularly at the interface between human systems and ecosystems as they relate to human survival, biocultural diversity, and productivity—require that today's students and teachers be exposed to diverse, interdisciplinary collaboration. This stimulates creativity, employs appropriate technology, integrates across cultural boundaries, examines economic relation-ships and processes, delves into environmental ethics, and practices engaged pedagogy in the community. Learning gardens, by their very nature, provide connectivity and partnership, not only among intra- and intergenerations of humans, but also foster partnership between interspecies, intercultural and intereconomic systems.

Attention to the following primary concepts guide Parajuli's (2002) develop-ment of sustainability curriculum:

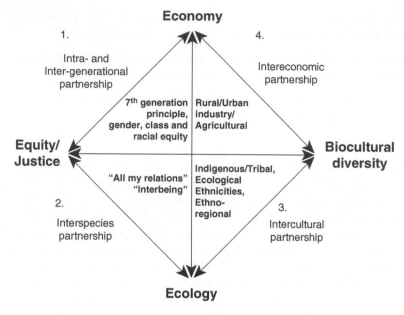

FIGURE 2.3 Parajuli's Partnership Model for Sustainability

1 *Intra- and intergenerational partnership*: Social class, gender, caste, race, ethnicity, and other human-created constructs, institutions, and practices of social inequities.

2 *Interspecies partnership*: Ecological, philosophical, and ethical aspects of human relationships with the more-than-human world, that is, the rest of the species.

3 *Intercultural partnership*: The field of biological, cultural, and linguistic diversity, diversity of knowledge systems and ways of knowing, teaching, and learning.

4 *Intereconomic partnership*: Social and economic institutions, arrangements of exchanges and surplus, fair trade and free trade between global North and global South, rural and urban relationships, agriculture and industry, producers, and consumers.

As a foundational curricular framework for sustainability, in the learning gardens students are engaged in an understanding within their communities on how culture and nature might thrive together in diversity. Students also learn to discern various paradigms and critique those that are unsustainable. Learning gardens curricula are interdisciplinary and weave science, math, social sciences, reading, and writing, as well as cultivate an ethic of care. Both the human and the more-than-human worlds serve as texts for learning. Building on the vibrant local and regional experiences, students also develop understandings that advance ecological and cultural awareness in tandem. Multicultural education thus gains new meaning in the learning gardens. The intersection of practice and theory, culture and nature, are also viscerally compelling and provide a fruitful line of intellectual cross-pollination cutting across disciplines and engaging fertile minds in the learning gardens (Parajuli, 2002; Williams, 2008).

Central to this discussion is the view that sustainability involves multi-directional and symbiotic relationships; therefore sustainability education is more than a descriptive education *about* sustainability and surpasses a prescriptive education *for* sustainability. Neither a deluge of information about social and ecological problems nor a dogmatic instruction toward a particular course of action constitutes effective sustainability education (Sterling, 2001; Stone, 2009). What is needed more than *description* or *prescription* is a reorientation of the guiding paradigm that shapes modern education. Sustainability itself consists of more than balancing divergent needs or contextualizing embedded systems; moreover it involves transforming consciousness from mechanistic toward ecological partnerships models. In this sense, the term *ecological* is used broadly to describe relationships between and among nature and culture. At all scales of living systems—from atomic to cosmic—relationships are a consistent theme around which life is organized (Capra, 2002). Thus, it is only natural to orient sustainability education around an ecological model of sustainability that facilitates and encourages diverse relationships. Learning gardens are one site where such co-evolutionary relationships are encouraged. Indeed, gardens represent rich sites of partnership among humans, plants, animals, pollinators, worms, and soil.

Gardens as Ecological Milieu

Permaculture originator Bill Mollison (1990) makes the thought-provoking observation that "all living things garden." Even ants cultivate and adjust their environment to better supply specific needs. Human-designed gardens are in this regard a natural response to meeting our needs for sustenance, security, and meaning. Inspired by an ecological perspective, we can endeavor to meet our human needs in partnership with the more-than-human world, in ways that are mutually beneficial to all. It is entirely possible to restructure our ways of being so as to come into congruence with the laws of the natural world and still reap abundance. Learning gardens are sites where such a holistic way of relating may be modeled in an educational atmosphere.

Filled with rigors of standardization and test taking, the institutional air of most schools is stagnant; the introduction of living things "bright and beautiful" offers a welcome relief. This alone is reason enough to plant gardens on school grounds. Practically speaking, in many urban districts, school grounds may be some of the only available arable land. Previously neglected or covered with sod or asphalt converted to gardens, these public lands can become assets as food producing sites, homes for wildlife, birds, and pollinators, and islands of beauty against the backdrop of gray concrete. The abundance of diversity seen in nature and manifested in gardens brings forth and fosters related cultural diversity. They make a positive contribution to student motivation, and development, as well as foster community hubs of organic activity organized at or around schools.

Gardens present an appropriate life-enriching ecological practice that guides curriculum, teaching, and learning. In an era characterized by educational malaise and apathy and amidst a repetitive discourse of racing to the top, gardens offer an alternative and regenerative model for bringing schools to life that differs significantly from mechanistic techno-scientific reform efforts oriented toward economic globalization.

Garden-Based Learning

When school gardens have strong roots in academic learning and they are integrated into school life, then they are not viewed as "add on" to the curriculum or as after-thoughts. With academic learning as a central goal, garden-based learning can be integrated into a continuum that addresses hitherto held dichotomies of nature and culture, school and community, ecology and economy, and life and learning. Garden-based learning is an instructional strategy that utilizes a garden as an instructional resource and teaching tool. "It encompasses programs, activities and projects in which the garden is the foundation for integrated learning, in and across disciplines, through active, engaging, and real-world experiences. In some settings, it is the educational curriculum and in others, it supports or enriches the curriculum" (Desmond, Grieshop, & Subramaniam, 2002, p. 7).

School Gardens: History, Research, and Recent Trends

Over the past 20 years, many educators have taken interest in using school grounds as a context for teaching and learning mediated through the life of gardens. This marks a return to an earlier movement in educational history when school gardens sprouted and flourished at school sites across the country (Hayden-Smith, 2006; Kohlstedt, 2008). As Subramaniam (2002) explains, youth gardening became a national and patriotic movement by the start of the twentieth century, when over one million students participated in growing food in school gardens. However, "the educational value of school gardens diminished and waned after World War I and their brief resurgence during World War II by the growing of Victory Gardens declined after 1944" (Subramaniam, 2002, p. 3).

According to historian Sally Gregory Kohlstedt (2008), in the late nineteenth century, progressive educators such as John Dewey proposed making the boundaries between school and society permeable, thus making life itself an educational enterprise. Natural environment and classroom learning were integrated. School gardens became highly visible and widespread examples of the living–learning connection. For progressive educators, school gardens served as tangible means to engage the young in education. "Gardens were encouraged in theory and in practice … in normal schools across the country" (Kohlstedt, 2008, p. 58). These gardens had a wide range of expectations and were flexible, based on context and locale.

With the various trends in educational reform defining the twentieth century, the school-garden movement ebbed and flowed. During World War I, with many farmers called away to war, school children were actually relied upon to help grow food for the nation. The federal government sponsored a national program called the United States School Garden Army which encouraged youth to garden for patriotic reasons (Hayden-Smith, 2006). This program, organized and delivered by the federal Bureau of Education, provided funding to promote school gardens as an educational initiative; it was also one of the first attempts to nationalize the curriculum (Trelstad, 1997). The gardens were school-based efforts, with strong connections to families and homes, and provided a curriculum that emphasized agricultural literacy to be integrated across the school curriculum. While the interest in school gardens gained momentum again during World War II to teach self-sufficiency and patriotism, in the latter half of the twentieth century, science and technology became the dominant metaphors (Damrow, 2005). With mass production of cheap food, large lawns became the defining features of affluence and culture. Growing one's own food was no longer seen as a necessity. Nor was learning to garden.

Since the early 1990s, however, school garden networks and green school initiatives have coalesced to support the creation of gardens across school districts where concrete school grounds are being de-paved for growing gardens. Hundreds of school gardens have been established in large and small districts spanning different bioregions. Garden curricula sprouting in districts, schools, and state education departments are designed for each grade level to meet subject standards, particularly for science, mathematics, language arts, nutrition, geography, literature, and health science, along with skills acquisition. Interest in meeting one or more of these standards has made the school garden an "instructional resource and tool" for "outdoor classrooms" (Dyment, 2005).

Multiple rationales, purposes, and outcomes are used to justify school gardens in the twenty-first century. In their synthesis of 48 empirical studies that were focused specifically on academic outcomes, Dilafruz Williams and Phillip Dixon (forthcoming) found that the research, while not extensive, has examined a variety of outcomes: academic learning in various subjects, changes in food habits, social development, attitudinal changes toward the environment, and school bonding, etc. Some of the studies (Koch, Waliczek, & Zajicek, 2006; Morgan, Warren, Lubans, Saunders, Quick, & Collins, 2010; Morris, Neustadter, & Zidenberg-Cherr, 2001) show that when students are directly involved in growing food in school gardens, they are more likely to try and taste new vegetables and appreciate the food that they have grown. Food gardens, in particular, are addressing issues of nutrition and health as Dorothy Blair (2009) and Emily Ozer (2007) have shown in their syntheses of research on garden-based learning. Garden-based learning is found to impact science learning outcomes as studied by Cynthia Klemmer, Tina Waliczek, and Jayne Zajicek (2005). In an evaluative study of a garden program with 277 third graders in Indiana, Amy Dirks and Kathryn Orvis (2005) were interested in how formal classroom settings using garden activities impacted science achievement and attitudes toward science, horticulture, and the environment. In pre-post measures, they found significant gains in knowledge and attitude toward science and agriculture.

In addition, studying the impact of a schoolyard habitat program in the Houston Independent School District, Phillip Danforth (2005) has shown that gardens can also positively affect math outcomes. In Rhode Island, Laura Castagnino (2005) found that an integrated garden-based science and writing curriculum is a useful tool in improving student vocabulary, observation skills, and the organization of written narratives. Thus, even writing is seen to be impacted positively where gardens are used by teachers to encourage expression by students, in this study. Another research study related to the involvement of non-white parents and their children in school gardens vis-à-vis growing food. In an experimental study of 52 second and third graders with majority Hispanic students in a school district in Texas, researchers Jacquelyn Alexander, Mary-Wales North, and Deborah Hendren (1995) found that participation in the gardening project had many positive effects. Students enjoyed gardening, had the chance to increase interactions with adults, and the program encouraged parents to seek out the school due to the garden. Students' academic curriculum was enhanced daily and they gained pleasure from watching the products of their garden labor flourish. Also, students increased their interactions with parents and other adults and the children learned the value of living things. The children spoke of the need for water and sunlight and commented that without these important ingredients their plants would die. The gardens were used both as a reward for hard work during the school day and as a supplement to the curriculum. The children learned to work as a team in order to complete all the necessary chores: weeding, watering, and fertilizing. They also learned the importance of caring for the gardens.

Furthermore, Michele Ratcliffe (2007), in developing a model of school gardens for teacher training, has found their impact at several levels: individual level, school level, and community/bioregional level. At the individual level, gardens are seen to affect academic achievement, and moral and social development in children, and also improvement in their health behaviors, as discussed earlier. At the school level, school gardens are also found to improve the overall ethos and learning environment. In addition, positive changes in schools and school grounds affect the community at large, according to Ratcliffe (2007).

Learning Gardens in Urban School Districts

To counter the manifest and literal lifelessness of schools and classrooms, an emerging national movement has been focusing on shaping the school grounds hitherto covered with asphalt or manicured lawns into green schoolyard habitats and school gardens across the globe. Between us, in the United States, the authors have visited over 80 school gardens in California, Colorado, Florida, Illinois, Oregon, and Texas. Take a journey with us to school gardens in four large urban school districts—Chicago, Denver, Portland, and San Francisco—with typical "urban" demographics: diversity in ethnicity/race, hundreds of languages, high rates of poverty, large numbers of English language learners, and students with a variety of special needs. Like many others, these districts have partnered with universities, government agencies, and non-profit organizations to seed, establish, and create gardens that engage students in learning about food, health, rainwater harvesting, native plants and animals, and/or wildlife. School grounds are birthing life and in doing so are livening education.

Chicago Public School District

Chicago Public School district, the nation's third largest school district, serving approximately 400,000 students, has a successful School Garden Initiative working to create school gardens and curriculum across the district. Launched in 2005, this ten-year initiative funds flower and vegetable gardens at schools helping students raise "everything from tomatoes to prairie grass." *The Chicago Garden Initiative: A Collaborative Model for Developing School Gardens that Work* (2008) provides a model for instituting outdoor learning landscapes on campuses to help children connect and interact with nature. The initiative is driven by a belief that gardens help students interact with nature, a relationship that enhances the overall mental and physical health of individuals. Over half of the almost 300 schools already have a garden. The collaborative includes Chicago Park District, the Chicago Botanic Garden, and the National Wildlife Foundation. We have found that garden-based initiatives often succeed when non-profit organizations, governmental agencies, and local businesses, join parents and education leaders to initiate and support these programs.

The Chicago High School of Agricultural Sciences maintains a farm, one of the last ones in the city of Chicago. In high demand, students learn about agricultural careers and also about taking care of farm animals. Several elementary schools, including Waters (K–8) and Tarkington (K–8) schools, have integrated gardens into their curriculum. In these school gardens, students grow, learn about, and also eat the food that they produce. At Waters Elementary, for instance, the "ecology" program covers field-based studies, recycling and conservation, and community gardens. Led by teacher and community activist Pete Leki, the school has an integrated approach to the school gardens which include both rainwater harvesting and growing food. In teaching students to conserve and reduce waste, the school also composts unfinished and waste lunch from the cafeteria, with students learning about making soil and then using it in their planting beds. During lunch break, students as young as first graders can be seen separating recyclable lunch waste and separating out uneaten food to be conserved and half-eaten fruits to be composted. Large compost bins have a prominent place in the garden inviting food and yard debris waste to be converted to soil, or, as Leki calls it, "black gold," for later use. He shares enthusiastically that the students are introduced to nature in a way that "fosters a personal connection" to the natural areas on site. Outdoor experiences are integrated into the classrooms in schools. Students are taught to keep field journals to write and draw about their experiences. Even the youngest students draw pictures or collect leaves for their journals. The community gardens on site also encourage the neighborhood to participate in and take an interest in public schools. According to the principal and teachers, the ecology program, which includes the garden program, is popular as students like being outdoors. Waters' ecology program was the winner of the 2003 Peggy Notebaert Nature Museum's Educational Leader Award.

Denver Public School District

In Denver Public Schools, with over 78,000 students, Denver Urban Gardens (DUG) and Slow Food Denver have worked in partnership to establish gardens at over 20 Denver Public Schools. DUG has initiated over a hundred community gardens, almost 20 of which are on school grounds. DUG offers an integrated nutrition and gardening curriculum to school communities, with a focus on supporting local classroom teachers in designing and weaving garden curriculum into various disciplines. The organization is particularly committed to helping schools that have predominantly large numbers of students from disadvantaged communities. Various subjects such as biology, ecology, horticulture, wellness, and nutrition are taught along with skills for recycling and composting. With support from DUG staff Judy Elliot and Jessica Romer, the teachers are able to integrate the curriculum, which is grade-level specific.

At Fairview Elementary (K–5) school, the garden's focus is on intergenerational relationships. There is a strong *Connecting Generations* model that enhances elder mentorship supported by bringing older adult volunteers to school grounds working side by side with young children in the gardens. The recent arrival of Somalian refugee communities housed within walking distance of the school has provided further opportunities to learn about land and growing food and also about their cultural culinary traditions. Even in the summer, when the school is closed, the garden is open, alive, and tended as several students and their teacher, along with DUG staff, continue to grow food and support the formation of community bonds. Fifth grade teacher Don Diehl has used the garden for several years as a resource for hands-on learning. His past fifth grade students return to the school in the summer to volunteer in the garden even after they have moved on to a middle school. Diehl's father also volunteers at the Fairview gardens; the school has large percentages of low-income and minority and refugee families. In the dreary and cold winter months, the gardens, with their murals, bring color and life to the school gardens.

At another school, Steele Elementary (K–5), the curriculum and pedagogy reflect a seasonal approach to learning about nutrition and gardening, both of which are woven into science. Slow Food Denver is a key partner in supporting the connections between gardens and the cafeteria. Enclosed within chain-link fences, painted signs reflect the herbs and vegetables flourishing in beds, some of which are raised, some are in pots, and most are in flat ground beds. Compost piles and bins are part of the garden environment. Slow Food Denver has partnered with the Denver Public Schools Nutrition Services to bring "scratch kitchen" to the schools. Andrew Novak, a chef by trade, has volunteered his time at the school to teach the students about nutrition and help them make the connection between the garden and the lunch plate. Similarly, at Eagleton Elementary (K–5), Slow Food's Gigia Kolouch provides a holistic nutrition program *Wellness in the Garden* that includes gardening, cooking, and market-related activities. This partnership program also includes support from Learning Landscapes at the University of Colorado. With a hands-on approach, the sea-sonally tuned program develops students' understanding not only of gardening, cooking, and nutrition but also of business related to youth markets. In particular, the gardens integrate science, and social and cultural studies. Besides produce, several gardens also have student artwork, thus bringing vitality to the school grounds. Additional robust garden programs at other schools across the city also have strong ongoing partnerships with parents and neighbors taking pride in their schools.

San Francisco Unified School District

San Francisco Unified School District (SFUSD), which serves approximately 50,000 students, has long partnered with San Francisco Green Schoolyards Alliance (SFGSA) and Urban Sprouts, among others, to create gardens at dozens of schools. According to SFGSA director Arlene Bucklin-Sporer and program manager Rachel Pringle (2010), the garden projects historically were funded entirely through parent association funds and private donations. However, SFGSA was able to leverage the 2003 and 2006 facilities upgrade school bond public funding to design and construct green schoolyards at 45 SFUSD elementary schools and 12 middle and high schools. At the first Chinese Immersion school in the country, Alice Fong Yu School (K–8), the parental community is actively involved in supporting culturally relevant curriculum centered on food.

A large plot of land at the school is transformed into gardens that are used by all grades at the school. Structured garden time is built into the school schedule. Art and craft are also integrated in the gardens. The hands-on experiential approach to gardening engages students in profound ways.

Another non-profit partner, Urban Sprouts, which is led by director Abby Jaramillo, grew out of a doctoral dissertation research conducted at Luther Burbank Middle School during the 2003–2004 academic school years. At the end of the study, teachers were excited about continuing the garden program and, according to Jaramillo, asked the research team to stay on and support them. In the first year, 100 youth worked in the school garden. This partnership with SFUSD has led to an expansion of school gardens to six middle and high schools. Since 2003, over 3,100 youth have benefited from Urban Sprouts serving as a staunch partner working with classroom teachers on hands-on garden learning with their students who grow and harvest vegetables and, as with Denver schools, bring produce to the cafeteria for salad bars.

At Martin Luther King Jr. Middle School, which is predominantly non-white, seventh and eighth graders have built several raised beds where salad greens are grown and students can easily access the food they grow. A shed and an outdoor kitchen have been constructed where students also learn the simple art of cooking. Students, who appear disengaged in the classroom, take active interest in gardens when outdoors. It is not unusual to see students with magnifying glasses examining soil and garden creatures. The partnership with Urban Sprouts has encouraged science teachers to do many curricular activities that include garden knowledge and skills both outside and inside the schools including experiments in their science laboratories. In 2009, a Farmers-in-Residence program was developed to involve culturally diverse families in the school garden and to enable them to grow their own food with their children right on school grounds. As is becoming common in other school districts, San Francisco schools are also addressing school meals that are healthier, fresher, and more sustainably grown.

Portland Public School District

Along with the Community Gardens program of the city, several non-profit organizations have been championing school gardens across Portland Public School (PPS) district, including Growing Gardens, Slow Food, Ecotrust, Master Gardeners, Hacienda, Waterworks, among others. As a result, there are at least 30 robust school gardens across the district. PPS serves 47,000 students with almost 100 languages represented at home. While not all schools call them learning gardens, a majority of these gardens are at elementary (grades K–5/K–8), and middle (grades 6–8) school sites. The few that engage high school students, such as Madison High School (grades 9–12), are planning career tracks in horticulture and landscape design. There is a *Garden of Wonders* at Abernethy

Elementary School (K–5), along with a scratch kitchen. The school, through the support of the Parent Teachers Association and grants, has been able to provide a garden coordinator who works closely with the teachers in integrating garden knowledge by grade level into the regular classroom. This structure of support is also seen at other schools such as Sunnyside Environmental School (K–8) and Lewis Elementary School (K–5). Growing Gardens, a non-profit organization has supported the start-up of gardens and also curricular efforts at high poverty schools such as Kelly Elementary (K–5) and Vernon Elementary (K–8), among others. Often the seeds of these partnerships emerge from parents, teachers, or community members interested in school gardens for a variety of reasons. In each case, we have found that garden supporters then join together to either seek monies or to work with their local neighborhoods and schools to involve and engage students in learning in the gardens.

Portland Public Schools Come to Life with Learning Gardens

Since 1995, author Dilafruz Williams has been involved in start-up and support of gardens in Portland Public Schools. She was also co-founder with Pramod Parajuli in 2003 of several food-based gardens in schools and later established the Learning Gardens Laboratory (LGL) where author Jonathan Brown taught for two years. Therefore, in this book, most of our examples are drawn from the network of learning gardens in Portland Public Schools with which we have had substantial and intimate connections.

Learning gardens in Portland, Oregon, that we have been involved with, have had one over-arching purpose: to advance academic achievement no matter in what zip code students and their families reside. A closely related goal has been to teach social justice while simultaneously building cultural competence and ecological literacy in tandem via the learning gardens, given the rich food and culinary traditions in the city. By bringing families who are either immigrants or refugees to work the land, the learning gardens encourage intercultural and intergenerational involvement.

In 2003, a partnership between Portland Public Schools and Portland State University (PSU) resulted in a Food-based Ecological Education Design (FEED) program which supported eight schools serving over three thousand students in elementary and middle schools. In addition, a Learning Gardens Laboratory was established in 2005 opposite Lane Middle School (grades 6–8) on a 12-acre site owned by the District and the City of Portland. In the learning gardens, students and families from a range of socioeconomic and culturally diverse backgrounds participate in learning how to grow food, build gardens, harvest, and cook. Students are simultaneously taught various integrated subjects. Dozens of teachers, parents, principals, and community members are participating in these gardens, most of which are at K–5 and K–8 schools and, recently, at a high school. The goals of the Learning Gardens program (Parajuli & Williams, 2005) are:

1   to foster multidisciplinary learning, connecting math, science, social sciences, languages, arts and aesthetics;
2   to promote multicultural learning representing multiple agricultural and culinary traditions of the parent community;
3   to cultivate intergenerational learning among young adults, parents and grandparents, educators, and others in the community;
4   to nurture multisensory learning by involving not only our heads but hands, hearts, skins, tongues, intestines, and palates.

We have involved students at Portland State University to do service-learning at these sites, thus ensuring another level of intergenerational partnerships for planning and design, construction, soil preparation, planting, maintenance, harvesting, documentation, and program outreach. The sites are designed and managed following permacultural principles using organic practices. Permaculture is a dynamic, holistic approach to gardening that emphasizes multiple uses of native plants, resource conservation, and community building. The courses at the university are integrated with a service-learning component for university students who work hand in hand with public school students. The learning gardens nurture students' love for nature, increase their understanding of the production and uses of edible and medicinal plants, educate about nutrition and the benefits of healthy eating habits, explore the cultural significance of foods including those from their native cultures, demonstrate the healing properties of nature and gardening, and elucidate the environmental benefits of local fresh foods and resource cycling.

Curriculum for Learning Gardens

Seasonal curriculum is developed and integrated with various subject standards. In food gardens, students not only grow their own produce, but also learn about the regional food economy by valuing, respecting, and eating food supplied by local farmers and suppliers who offer seasonal and sustainably grown produce that has been highlighted and promoted by the District's Nutrition Services Director, who champions "harvest of the month" and offers local salad bars in the cafeteria. The "Harvest of the Month" program increases students' awareness of and interest in seasonal and regional fruits and vegetables by integrating fresh food in school lunch with classroom activities.

In rain gardens and native gardens, students learn other sustainable practices such as rainwater harvesting, landscaping with native plants, and incorporating wildlife into gardens. Curricular experiences between classroom learning and what is learned in the gardens are integrated. Table 2.1 displays a sample of the

TABLE 2.1 Curricular goals and subject integration

Grade and Subject	Benchmark	Lessons and activities in the learning gardens—topics
6th Grade		
Life Science	Group and classify organisms based on a variety of characteristics	Corn and Family
Mathematics	Explore two-dimensional geometry	Designing a Square-Foot Garden
Social Studies	Know that societies influence what aspects of technology are developed and how these are used. People control technology and are responsible for its effects	Native American Culture and Native Teas
7th Grade		
Physical Science	Understand that chemical or physical changes can be explained by changes in the arrangement and motion of atoms and molecules	The Chemistry of Corn
Mathematics	Analyze and graph data	Soil Investigation
Science Inquiry	Write a set of logical, practical, safe, and ethical steps which addresses the question or hypothesis	Making Your Own Potting Soil
8th Grade		
Earth Science	Identify factors affecting water flow, soil erosion, and deposition	The Learning Garden in the Watershed
Mathematics	Explore Algebra, estimation, and prediction	Planning the Cafeteria Garden
Health	Explain the importance of variety and moderation in food selection and consumption	Designing the Cafeteria Garden
Social Studies	Understand how the process of urbanization affects the physical environment of a place	Studying Local Places

curricular goals and integration of subjects at the Learning Gardens Laboratory in Portland, Oregon (adapted from Dardis, Parajuli, & Williams, 2008).

The curricular goals are integrated with garden activities and learning for:

- Science, Mathematics, Language Arts and Social Studies content of each grade

- Gardening skills and knowledge appropriate for each grade
- Progressive personal and social development of the student over the three years
- Seasonal cycles of the garden and the natural environment
- Cyclical nature of community activities of the school and neighborhood
- Cultural foundations and traditions of the diverse student body and of the Native peoples of the region.

In the process of learning from the gardens, children and youth begin to appreciate ways in which the health of the individuals, the health of the land, and the health of their communities are intertwined.

Integrated and Seasonal Curriculum

Growing food, harvesting, and eating are perhaps the most primal, intimate and sensuous ways we connect with the rest of nature, since it is through food that humanity's most intimate and essential connections to the soil, earth, and other creatures are expressed and consummated (Nabhan 1997; Capra 2002; Cobb, 1969; Kiefer & Kemple, 1998). The sequestering, growing, and trading of food have shaped all societies historically and culturally throughout time. As practicing educators in the field, we also know that gardens and food strike a visceral chord in the classroom, probably more than any other stimulus, particularly when there is lack of attention, interest, self-efficacy, or focus. Combined with these basic perspectives on food is the sharp new focus in many school districts across the United States on health and nutrition. For us, *ecology* represents the systems, processes, and relationships of the natural world. These are typically explored in the curricular goals of life sciences, physical sciences, and earth sciences within school curricula but they also undergird the everyday lives of children as in our taken-for-granted weather and seasonal cycles or the pull of the moon. The closest children can get to appreciate ecology is through their tactile, sensorial, and bodily experiences and relationships with air, water, food, and soil. Since the health of children is directly related to the quality of their natural environment, we envision schools participating in bringing life to learning even as learning comes to life *via* school gardens. The learning gardens provide comprehensive, experiential, and transformative experiences to students creating a grounded sense of place before they can truly become stewards of the land. As they design their own garden beds, students begin to fit plants and life cycles into the context of place. A significant part of each year's curriculum involves students exploring and mapping the gardens within the contexts of ever-increasing geographical areas, from the garden itself in fifth grade, to the neighborhood in sixth grade, to the city in seventh grade and watershed in eighth grade. Students' sense of place develops as they see how the garden compares to land-uses at home and elsewhere in their community. Interaction with community groups helps to further define their place within the bigger society. Studies of maps, satellite photographs and field trips then advance their understanding of place within the geography of their city and the bioregion.

Learning gardens lessons are designed with the same philosophy as is a good permaculture farm founded on the principles of an interdisciplinary approach to building soil while creating food production systems in a regenerative way. Formulated by Australians Bill Mollison (1990) and David Holmgren (2002), permaculture principles have been developed from protracted observation of natural systems and time-tested indigenous land-use methods. We suggest that insights from permaculture can guide our approach to working with, rather than against, natural systems, and can be applied to educational systems as well. For example, the permaculture principles "integrate rather than separate" and "value diversity" are commonly applied to the planting of symbiotic species in a garden to ensure productivity and health. We can similarly integrate symbiotic, or mutually supportive, diverse ideas in educational practice. Such a "guild of ideas" could address multiple learning styles and perspectives and would contrast a "monoculture of the mind." As another example, permaculture teaches that the "edge" between two systems is often vibrant due to the formation of particular microclimates and sharing of diverse resources. The edge where a pond meets the land, for instance, is home to species that require a little of both systems. We can enhance the "edge" effect in education through creating permeable school walls and encouraging greater contact with the surrounding social, ecological, and cultural systems. In learning gardens "inputs and outputs" are integrated tightly. Through recycling and composting, for example, experiences cycle back into the curriculum as if it were soil. Using a technique known as "stacking functions," we can secure higher academic results for our efforts and simultaneously inculcate healthy food habits that can also enhance the health of our youth, children, and communities. Accordingly, in the design of the learning gardens, we propose the adoption of permaculture and whole-systems principles such that soil, water, sunlight, biomass, plants, animals, people, and sustainable technologies, are connected to one another.

Student Reflections in the Learning Gardens

Learning in the garden takes on tangible, pragmatic, and embodied meaning. An analysis of student artwork and free writing indicates that students are learning principles of systems and holistic thinking. Students' writing captures this quality of insight:

> I sit here on my clump of grass. I watch the beautiful rhythm of the constant turning and weaving of the patterns in the shimmering water.

My pants become a sponge and absorb all the moisture around. I start to tune into the whispering hum of the swarming mosquitoes circling my head. The usual breeze seems much more extreme undisturbed by the common heavy machines.

I look up high to find a nice blue sky. I squint my eyes protecting me from the bright light coming from behind the white fluff clouds. But then I focus in on manmade power lines closer than far. I realize that I'm not in paradise.

The feeling has gone.

(7th grade student)

The above meditative journal entry communicates the depth of awareness students are able to cultivate in quiet contemplation in learning gardens. Such thoughtful observation and careful attunement to the diversity of sensorial phenomenon is hardly possible in conventional school settings alternately filled with ringing bells and bustling movement or stagnant air and empty halls flooded with fluorescent light. For a moment, this student grasps insight into the wealth of living phenomena that too often lie hidden in plain sight. Similarly, in the poem below, a student encapsulates nuanced ecological and cultural understandings of stinging nettles in brilliantly rhymed couplets. What to many is a pernicious weed or an obnoxious garden nuisance has become a valued friend to this student poet.

Stinging Nettles

From a seed I start off slow
Add some water and light I start to grow
I'm skinny and I'm very tall
Stinging Nettle is what I'm called
I grow in a place called JEAN's farm
My hobbies are stinging arms
I'm made into rope and tea
It's easy to see why people love me!
I have many more uses I can't count them all
And when you're with me you'll have a ball
I'll die out just as you feared
But don't worry I'll come back in the spring next year.
What I have to say before I go
This is something you should always know
You must remember just one thing
Get too close and then I'll sting

(7th grade student)

In our site visits to school gardens, we have found students from different socio-economic backgrounds involved in service-learning. They harvest food and take it to the food bank, feed the homeless and the hungry, and participate at projects in shelters. These experiences draw awareness to the large percentages of children for whom free public school meals are sometimes the only meals they receive in a day. Students are exposed to an understanding of how in an affluent society, there still are people who have no food or shelter. They begin to think critically about the importance of learning to responsibly grow and share food locally, in light of energy and transportation problems impacting all communities. One third grade student shares her thoughts when asked why the local is important to her: "Local is important for many reasons, but I'll just list three: It doesn't waste fuel to get food here. Food is a lot fresher. And, it is good to support local businesses." Growing food in learning gardens enables students to perceive many connections linking social and natural systems, as another third grade student reasons about why paying attention to where food comes from is important: "because when, for example, a box of tomatoes reads, 'shipped from China' they may have additives in them to make them last longer. But local stuff usually doesn't have many additives because it comes locally." An 8-year-old boy explains: "A garden helps the environment, gives animals a home, you can grow your own fruit and vegetables and it just looks nice"; while a girl in his class states, a garden "brings people together to make a place where you can get food and be calm." The above comments are in keeping with Fritjof Capra's "web of relationships," and the indigenous understanding of "all my relations." Engaged in the vibrant living systems of learning gardens, students are able to perceive the interconnections between seemingly disparate sustainability concerns such as hunger, food and transportation costs, value of local food, and gardens; moreover, they recognize that any one of these problems or solutions cannot be understood in isolation, that each is embedded within one another, connected in webs of relationships.

Student writings and comments presented here are but a small sample of the innumerable essays, poems, drawings, and interviews that are indicators of an ethic and practice of care for self, land, and community that are developed from relationships with a place over time. This care is important for investing in the long-term health and vitality of self, land, and community. The poetic links between ecological communities and human communities emerge as no living being is seen in isolation. Gardens are also vibrant sites for engaging students in critical thinking related to environmental and food disparities and social justice. For instance, students at one of the middle schools harvest food and take it to the food bank, feed the homeless and the hungry, and participate at projects in shelters. No life—whether that of stinging nettles, or a spider, a *nasturtium*, a tree, or a homeless and hungry man—goes unnoticed. Relationships are recognized by children of all ages.

Indirect Academic Outcomes

There are a number of less conspicuous and indirect learning outcomes associated with school gardens. One example is the reawakening of children's sense of wonder and awe at the magnificence of life present in gardens. Paul Krapfel (1999) has described ways in which gardens stimulate children's curiosity and imagination and deepen environmental participation. The adoption of secret garden spots helps children to get quiet with themselves and observe nature within the bounds of the busy bustle of school life (Young, 2001). Gardens provide a safe space for repose, which is important for children given the diminishment of opportunities for young people to access natural areas for such reflective and exploratory time (Louv, 2008). The slow pace of life in the garden contrasts the dominant social norms of speed and instant gratification codified as much by video games such as "Race to the Top." Play in the garden context is "unplugged" and thereby also contrasts the dominant norms of modern childhood. Together, these learning outcomes are critical, as wonder, curiosity, imagination, and play are becoming endangered experiences of childhood.

Learning gardens can foreground the cultural and historic dimensions of food, land, and growing practices. Food is intimately connected with land and culture in significant ways; the variations in landscapes have given rise to different food-producing capacities and thereby diverse cultural uses of food. For example, students can consider the central role of corn in the mythic traditions of indigenous people in North America as well as in the modern diet. First domesticated from the wild plant *teosinte* near Oaxaca, Mexico, corn has been critical to both the dietary and the cultural constitution of indigenous people for over 10,000 years (Sachs, 1996). The Spanish phrase, "*Estoy caminando maíz,*" or "I am corn walking," refers both to the centrality of corn in the Mexican diet and to the spiritual connection between people and corn. Today, corn is ubiquitous in processed food, central to the modern diet in a distinctly different way.

School learning gardens also provide opportunities to celebrate cultural connections to food through the creation of historical or culturally focused plant guilds. The "Three Sisters Garden" is a plant guild of the three sisters—bean, corn, and squash—with historical and cultural connections to indigenous spiritual and land practices in the northeastern part of North America, ranging from the Ohio River valley, northeast into northern Vermont. Such intentional planting in a school learning garden accomplishes multiple tasks at once: it grows corn, beans, and squash together in a small area, provides a tangible lesson on indigenous land-use practices and spirituality, and offers an example of sustainable ecological design using "companion planting."

Many urban schools are located within what are known as "food deserts," defined as areas of the city with insufficient access to grocery stores and healthy fresh food. Because grocery stores base their location upon predicted sales and revenue projections, supermarket chains often avoid low-income areas

(Winne, 2008). The lack of conveniently located grocery stores in low-income areas is a contributing factor to class and racial stratification of sufficient food access; as many inner-city residents do not own cars, transportation to distant food stores can be a hardship. Furthermore, variation in diet and consumption of fresh produce is associated with higher food costs. Hence, learning gardens in schools are also used as venues for addressing issues of food insecurity and community hunger.

The constellation of increased distance to grocery stores, transportation obstacles, and higher costs for fresh food all contribute to food insecurity among low-income school-aged children, in particular. Inadequate nutrition is closely linked with school absences, fatigue, concentration problems, anxiety, depression, behavior issues, diabetes, and infectious disease. While school gardens cannot ameliorate all of these complex issues directly, growing fresh food on the school grounds brings this often silent issue of hunger and food insecurity to the table as a topic of concern. School garden projects in several urban districts such as San Francisco, Portland, Boston, Denver, and Houston donate harvested produce to local food banks, or engage students in serving at soup kitchens that help the homeless and the hungry. These activities encourage an ethic of community empowerment and person-to-person connectedness, as students realize their ability to positively contribute to partial resolution of intractable social problems such as food access and hunger. Where the supermarket chains will not locate on account of subprime sales projections, school gardens can flourish as vibrant and healthy small-scale community hubs.

Learning gardens can also be designed as sites encouraging "nature in the neighborhood." Through activities such as rainwater harvesting, planting of butterfly and other wildlife guilds, and providing insect and wildlife habitats, school gardens can teach children the value of connecting with and offering human help to biotic neighbors. These activities can be as simple as piling up rocks and sticks at the edge of cultivated garden spaces or planting white clover in pathways.

A function of school gardens directly related to providing wildlife habitat is closing the food cycle in the school community. Through composting, children are taking responsibility for food waste commonly sent to landfills and are able to create an intergenerational gift to soil. The soil at many school grounds is nutrient deficient as a result of long neglect; hence compost is needed at almost all sites. Rather than an added expense, this need presents a golden opportunity for schools to harvest abundant cafeteria waste. Furthermore, composting fits perfectly with the academic school year, which is somewhat out of sync with the summer gardening season. The fading of summer abundance yields a rich harvest of withered stalks and plant matter, which when combined with food waste, can create fertile compost over winter. For older students, a school waste audit can be appropriate, while for younger children an appreciation of decomposers as contributors to the web of life may be sufficient.

Designing School Gardens in Difficult Locations

In some locations, school gardens may not be possible to build due to lack of arable land, lack of time, or polluted soil. In these less than ideal conditions, broad-scale techniques for creative solutions are sought. Creative use of trellises and vertical gardening can help schools with limited space optimize small growing areas. Spiral garden design, keyhole beds, and intensive planting methods are more examples of applicable techniques for gardens located in less than ideal locations. One solution for building soil in such instances can be found in worm bins, which are low-cost and easy to build. Worm bins offer one way to begin food scrap recycling inside classrooms, an activity that simultaneously closes the nutrient loop by recycling food scraps and provides an integrative context for thinking about invertebrate life, nutrient cycles, temperature, moisture, waste as food, and so forth. Another possibility that we are excited about is straw bale mulch gardening, an experimental permaculture technique used in a number of places worldwide. Straw bale mulch gardening addresses the challenges associated with sites built with concrete or characterized by polluted soil. To implement this method, straw bales are stood on end so that the grain of the straw is vertical. To this surface, nitrogen rich material such as composted steer manure or bonemeal is applied in a thick mat. Over several months, this nitrogen mat is watered into the straw bales, which in turn begin to decompose on the inside while maintaining their exterior form. Plants can be grown in the top of the partially decomposed straw bales stacked atop of concrete or polluted soil with little worry of contamination. The bales hold their form for several years, elevating the garden space a few feet above the ground. This method may be particularly appropriate for small children who may otherwise walk on in-ground garden beds. More traditional wooden framed raised beds can also be used atop asphalt surfaces or over contaminated soil as long as an impervious barrier such as plastic is applied.

Intersection of Culture and Ecology in Learning Gardens

Ethnobiologist and nature writer Gary Paul Nabhan once remarked that if he had to choose the best ambassadors of biodiversity, he would select storytellers, musicians, and herbalists. Along with others such as Wendell Berry and Barry Lopez, Nabhan (1997) identifies and addresses problems arising from our accelerating extinction of experience, extinction of meaning, and extinction of relationships. He explains that we are experiencing not only extinction of species, but also an extinction of ecological interactions (Nabhan, 1997). In our work in traditional classrooms, we have observed that reduction of ecological interactions directly impinges upon students' ability to relate with one another as human beings and with the more-than-human world. Learning gardens are potential sites for inquiry and restoration of experience, meaning, and relationships, which can reaffirm the threads of reciprocity, embodiment, interaction, networks of meanings, and experiences that students have with the living soil and earth.

Educators at all levels recognize that schools and classrooms have become and are continuing to become more and more culturally diverse. Yet teachers often find it difficult to relate to students' different ethnicities and cultures. Schools often embody biased monocultural learning activities that alienate rather than include students. Learning gardens provide additional pathways to education that particularly support diverse learning styles and interests as well as diverse cultural understandings. The basic questioning and linking of human thought and culture, both past and present, posed at this early stage in their learning, can awaken young students' imagination and at the same time foster intellectual discernment needed in contemporary life.

Large urban school districts are struggling to close the academic achievement gap especially among low-income and ethnic minority students. Immigrants and refugees, in particular, often feel disengaged from the educational and cultural realm of the large cities. Schools need tools to address diversity and engage students and their families. Learning gardens pedagogy organized around food gardens draws students into a mode of inquiry related to the food and culinary traditions of diverse families and communities. At the Learning Gardens Laboratory in Portland, for example, a multicultural family garden was initiated honoring immigrant and refugee families and parents to engage them in food-based educational activities in schools as well as in gardens. These family gardens support the exploration of multiculturalism, historical, cultural, and philosophical perspectives on the interconnections between people, ecosystems, and the ecology of their food (Berry, 2004; Lopez, 1978; Maffi, 2005; Nabhan, 1997; Posey, 1999). As founder Parajuli (2006) explains: Children and their parents delve into questions such as: What kinds of environments did our ancestors come from? What did they eat there? What did they grow and produce in their respective environments? What foods and traditions did they bring with them to the United States? What did they leave behind when they migrated? How does food acculturate us to our new environment? These questions are foundational to providing a basic, grounded journey into the education of young students in multicultural learning and living environments that often make up marginalized neighborhoods of urban areas. These profound questions concern people's relationships with nature, culture, and sustenance of both.

We believe that these lines of inquiry have potential to alter students' self-concept in relation to others as we seek to redefine education and sustainability. This contact does more than improve dietary health and daily exercise.

Rooted in Soil

Through learning gardens, we are seeking to help students to reflect on the fundamental questions of what it means to be human. In the process of learning from the gardens, children and youth begin to appreciate ways in which the health of the individuals, the health of the land, and the health of their communities are intertwined. Learning gardens on school grounds provide diverse and rich texts for nurturing students' connection with the more-than-human world. Direct engagement with life in the gardens brings children into contact with a richly diverse biotic world of living soil that is too often "out of sight" and "out of mind." Much more than a fanciful educational trend, learning gardens challenge mechanistic perceptions of living systems as complex machines and remind us of the interconnectedness of all life. Beyond the blossoms and bountiful harvest of the gardens themselves, though, lies the hidden living soil that sustains the entire system.

3
LIVING SOIL AS METAPHORICAL CONSTRUCT FOR EDUCATION

The beauty and the abundant harvest of learning gardens have captured our attention as dynamic venues for education. When we walk down soft woodchip paths of learning gardens, we revel in the abundant harvest of fresh vegetables and fruit, admire the splendid colors of summer flowers or autumn leaves, and rejoice in the happy commotion of students bustling from one learning station to another. Underneath our feet lies the apparently quiet soil which breathes life into the delicious culinary harvest, animates spectacular flora, and supports the educational programs of school learning gardens.

Without soil, there can be no life. This simple fact remains true no matter where we are; we live a terrestrial existence reliant upon living soil. Regardless of what city, state, or school district we find ourselves in, we can safely return to soil as the vibrant source of material wealth. Through practicing organic techniques such as adding compost, school gardens are actively contributing to the enrichment of our planet's soils, which are too often degraded or paved in the name of progress. Long polluted by chemicals and heavy metals, and in many communities buried beneath masses of concrete or asphalt, living soil is far from commonplace. It is out of mind since it is often buried under asphalt. In learning gardens, we find the opposite to be true, as students and their teachers gracefully "unearth," nurture, and sustain the infinite source of eternal life right beneath our feet, as living soil is brought back "in sight and in mind."

For children, learning gardens provide an excellent place to commune with life. In particular, the living soil of learning gardens connects children with life below the feet. We celebrate learning gardens as dynamic and creative sites for nurturing ecological and cultural literacy. Indeed, students are able to develop intimate connection with plants, animals, insects, wind, and water through awakening their curiosity, wonder, critical thinking skills, and imagination in commune with soil directly and indirectly. In this chapter, we seek to leverage interest in learning gardens as a whole-system design solution addressing interrelated social, ecological, and educational problems, and dig deeper into the living soil that literally and metaphorically supports learning gardens produce and pedagogy.

Living Soil as Metaphor

In moving the emergent sustainability education discourse beyond the trappings of the modernistic metaphors and worldviews described in Chapter 1, we propose the development of a regenerative metaphorical language to inform sustainability teaching and learning. Both of us have initiated and cultivated school learning gardens, which offer a promising avenue and entry point toward engaging students of all ages in learning about sustainability in a hands-on practical manner. Through our mutual and individual experiences with learning gardens, we have found *living soil* to be a potent metaphorical lens through which to begin a formulation of ecologically grounded principles for sustainability education. We recognize that this framework is somewhat limited as it is explicitly terrestrial. Nor do we claim to be soil experts. However restricted, our intention is to begin an iterative process to encourage exploration of ecologically grounded phenomena such as soil as the theoretical basis for sustainability education, rather than recycling inflexible mechanistic metaphors and their corresponding cultural assumptions.

Fritjof Capra (2002) suggests one way to break ourselves from returning to the machine metaphor is through development of new paradigms. He suggests reframing social organizations as living organisms rather than machines as a first step. Embracing this approach, we look to the soil of learning gardens in developing a context-specific ecological root metaphor as an alternative to mechanistic metaphors that guide education at present. Observing the necessity of soil in sustaining the varied and abundant harvest of produce that grows in learning gardens, we here position it as a new metaphor for guiding learning gardens design and pedagogy. Context is critical as it links metaphoric understandings to the tangible reality of everyday phenomena. Indeed, it is the ubiquity of machines that makes mechanistic thinking so "second-nature." In transitioning away from the machine and toward living soil as a guiding metaphor, we need to cultivate similar understanding and relationship with soil, given that it is a far less common component of modern experience. Negative connotation of soil as "dirty" and "low-status" along with physical abuse and disguise of soil under parking lots and AstroTurf playing fields make living soil somewhat an unknown entity and thus an awkward basis for a new root metaphor.

Soil is the living skin of the earth. A living soil is far more than a mere growing medium; it is a confluence of eroding rock, decomposing biomass, micro-organisms, animals, insects, water, and air. Complex food webs and interdependencies constantly remake soil as it passes through the digestive tracts of worms and various soil micro-organisms. A living soil is layered, develops over time, is fragile, and contains and supports life. Moreover, soil actively *promotes life and is itself alive*. If we are not careful, centuries of natural soil building work can be undone in a single year through wind, water erosion, or human activity. Living soil holds incredible amounts of water, and is a key component of the water cycle. Soil rich in cured organic matter, or *humus*, is able to absorb several times its weight in water, a characteristic which reduces runoff in flood times and improves plant growth during drought times.

Through our lived experience of growing food for many years we have each come to learn that soils have "personalities," and that successful interaction depends in large part upon a healthy relationship with the soil, a factor that is overlooked in a merely biophysical/chemical description. Thinking of soil as a malleable complex growing medium biases perception and action toward anthropocentric needs and wants. Viewing soil as a living entity suggests a radically different relationship. We can choose to enter into a mutually supportive relationship with living soil and "get to know" soils. A key step is learning ways to grow soil.

There are numerous established techniques that can guide efforts to contribute positively to soil health. We can study and apply strategies from biointensive gardening (Jeavons, 1974), no-till polycultural farming (Fukuoka, 1978), and composting in place (Hemenway, 2000; Mollison, 1990) as initial approaches toward returning life to soil. The methods mentioned above have been tested and documented as effective approaches to soil health building over more than 30 years. Moreover, these techniques require no chemical additives or harsh industrial processes; rather, they rely upon the clean renewable energy of human bodies and minds, are safe for participants of all ages, and are pleasant and fun. Furthermore, the described techniques set in motion powerful positive feedback loops. For example, by returning large volumes of biomass to soil through sheet composting in place, the soil is able to produce even more prodigious quantities of biomass, thus providing its own organic fertilizer for the next growing season.

The simple soil building methods described above especially contribute to the formation of *humus* within soil. *Humus* is the hallmark of living soils. Land that is rich in humus is able to hold vast quantities of water, support a diverse community of bacteria, worms, and micro-organisms, and yield an abundance of vegetables and fruits. Like soil, *humus* is slowly formed and can be quickly exhausted. Continual replacement and judicious use (e.g. allowing for fallow periods) complement initial soil building techniques and ensure long lasting living soils.

In school learning gardens, students learn these and other ways to enrich soil and in so doing contribute to bringing forth life. For example, steaming compost heaps demonstrate the possibility of recycling decomposing matter into rich plant food, which in turn gives life to more human food. The simple lessons of the compost heap teach children that there is no throwing "away" and that waste is a condition of overproduction or underutilization of a useful resource. Instead of just discarding apple cores or autumn leaves as unwanted "waste," students see that they can transform debris into fertile compost as a way to actively enrich life.

Diverse spiritual teachings also remind us that we arise from the soil and to the soil eventually return. In terms each of us can strive to understand, our daily bread reaches us not by providence but through the nurturing interaction of people and soil. In modern industrial food systems, this basic fact is obscured from view, yet it remains true. Soil conditions eventually translate into consumed food calories. Healthy food grows from healthy soil. Though scientists have made advances in fields of genetic engineering and hydroponics, it is simply unfeasible to grow orchards in liquid nutrient solution, besides which, even the most perfected laboratory seed must eventually find sufficient moisture and warmth to germinate in a terrestrial soil in a specific place.

It is of utmost importance that we seek to regain a tangible understanding of soil and create favorable conditions for students to engage with and be engaged by living soil. In urban school districts, soil is too often paved over by asphalt or concrete. In rural districts, soil may be polluted by agricultural chemicals. Ironically, in many communities, school ground is the only available land not claimed by development. We can transform this public land into abundant and educative school gardens. Learning gardens on school grounds are one practical way to create opportunities for such soil learning.

But, planting gardens alone is not enough. More important than redesigning the school landscape from sod to gardens is to redesign the mindscape: from domination to compassion, from competition to cooperation, and from mechanical relationships to ecological relationships. We cannot go far if we are simply moving plants around a garden as unrelated objects subject to human control. Similarly, we cannot affect change in education through shifting the venue from indoor to outdoor without an associated shift in guiding mental models, for even as we unearth soil in gardens we may unintentionally reinforce mechanistic metaphors and marginalize living systems. As an example, when we describe photosynthesis with the metaphorical language of the factory, we reassert the primacy of the machine as the guiding point of reference. Thus, as we cultivate school grounds, we must also cultivate a new metaphoric consciousness in our minds. As Bowers (1997) asserts, mechanistic root metaphors are so deeply ingrained in the collective psyche that they influence education in numerous unseen ways. Based on this insight we caution that, lacking an alternative regenerative life affirming paradigm, learning gardens design and pedagogy can unwittingly reinforce taken-for-granted mechanistic metaphors and a general domination mindset. For example, school gardens can too easily become sites for further perpetuation of conventional gender roles and cultural reproduction.

Growing straight rows of monocultures of plants conveys in physical manifestation a monoculture of the mind (Shiva, 1993) that overlooks the critical role of diversity in biological and cultural systems. We believe that connecting with the soil of learning gardens can help us to transcend the limitations of the mechanistic metaphorical orientation and show us the way toward an alternative paradigmatic framework.

Living Soil, Schools, and Learning Gardens

The soil in learning gardens can be used in designing sustainability programs and pedagogy, and can also be extended as the basis of an ecological metaphor for thinking about bringing life to schools in general. In direct contrast to the mechanistic and industrial images routinely used to frame education in general, we propose living soil as a metaphorical lens to guide teaching and learning. As an explicitly ecological metaphor, living soil implies the necessity and value of life and requires an attitude of mutual reciprocity. Non-violence is a key here. In gardening, we cannot go far without aiding the soil with respectful attitude and action. Such qualities of respect and reciprocity are often absent in mechanistic and industrial metaphors that characterize schools as elaborate machines. Dominant metaphors referencing schools as "knowledge factories" fail to incorporate life into schools, and are thus ineffective in bringing schools to life. Likewise, reform metaphors comparing teaching and learning to a competitive "race" suggest an atmosphere in which mutual reciprocity or cooperation has little significance. Living soil is neither a factory nor a race. It is the ever-changing biocultural milieu in which seeds are able to germinate, grow, change, and express themselves. It is composed of past lives—plant, animal, bacterial—and gives rise to new life.

Living soil is both a literal and a metaphorical medium for germinating and nurturing "seeds of change." No matter if the "seed" in question is genetically modified in a laboratory, passed down from generations in a mason jar, or the kernel of a transformative idea, eventually all seeds will necessarily take root and flourish in specific place-based terrestrial "soil." The conditions "on the ground" are thus of interest in discussing change movements. We take pleasure both in digging the literal living soil of learning gardens and in encouraging similar living vibrancy in the organizational and pedagogical constitution of schools. In doing so, we move away from the metaphor of the machine and promote a vision of schools as a complex interdependent living medium/organism we call "soil."

Healthy plants can only thrive in a vibrant living soil; we further assert that innovative educational initiatives can likewise only thrive in a rich, diverse, and living metaphorical "soil," or context. A living soil is a dynamic ever-changing complex ecosystem which cannot be simplified or reduced to its component parts. With this metaphorical framework, we now present seven attributes that contribute to an ecological vision to bring life to schools. Through shifting the underlying metaphors toward living soil, we hope to encourage respect for the diversity of human and biotic communities, reawaken dormant sensory systems, alert students to their position within the web of life, ground learning in place, and introduce alternative regenerative ways of living. Learning gardens are critical venues for learning the lifeways of soil and unlocking the mysteries of education. As gardens are physically based in soil and located on school grounds, they are sites uniquely situated to engage students in unearthing the wisdom of soil. Through experience in learning gardens we begin the slow process of bringing living soil back "in sight and in mind." To explore this, we deepen our reading of living soil as a metaphor by highlighting seven attributes of soil that we have observed in the gardens that can enrich pedagogy.

Learning Gardens Principles: Linking Pedagogy and Pedology

As seen in Figure 3.1, the following seven attributes emerge from the central metaphor of *living soil* that guide learning gardens pedagogy: cultivating a sense of place, fostering curiosity and wonder, discovering rhythm and scale, valuing

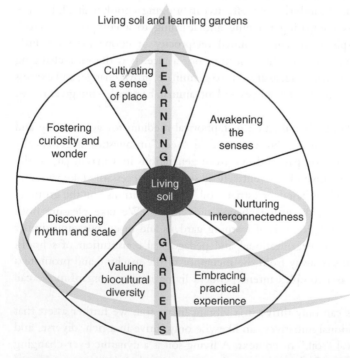

Living soil and learning gardens

FIGURE 3.1 Living Soil and Learning Gardens: Principles Linking Pedagogy and Pedology

biocultural diversity, embracing practical experience, nuturing interconnectedness, and awakening the senses. At the core of the figure is the soil from which emerge principles of pedagogy for designing education. In serving as guiding principles of pedagogy, the attributes are woven into the framework of learning gardens as can be seen in the spiral in the figure. The seven principles of pedagogy remain anchored and integrated in learning gardens.

1   *Cultivating a sense of place.* Living soil is inherently local. With regard to the interconnectedness of all things, it is necessary to consider soil within its ecological and bioregional context. We begin where we are, working with the soil directly on school grounds. At the same time, we broaden our understanding of the world around us in successive waves of relationship, moving in series from the school grounds to the neighborhood, to the community level, to the bioregional, and finally the global level. In this way we see both the local and the interconnected nature of social, ecological, and economic actions within a globalized world. Conventionally, education is too often framed with little regard to the places in which schools actually exist. Though schools exist in physical communities, filled with social and biotic actors and relationships, education follows a de-contextualized script (curriculum) without reference to the local context. A focus on the global overlooks the relevance of the local and the intricate interconnections between the local and the global. Realizing that it is impossible to "think globally," and rejecting the simplicity of "think locally" as an alternative, we propose "think ecologically, act locally" (Williams, 1995) as a regenerative possibility. Thinking ecologically involves understanding hidden connections between phenomena, whether biological, social, or cultural. Acting locally refers to applying ecological insights within the context of place.

2   *Fostering curiosity and wonder.* A living soil invites questioning. Endless questioning. How do worms breathe underground? How is *humus* formed? How does water move through the living soil? A living education likewise encourages curiosity and wonder, and invites endless questions. Through questioning, children seek to make meaning of the unknown. It is unnecessary that children's latent urge to ask questions be stifled by standardized tests and the "get it right" syndrome of modern education. Children's uninhibited curiosity can surface taken-for-granted assumptions and inadequate understandings of phenomena—two types of insights that will be deeply significant in transitioning toward sustainability. As Carson (1956) explains, to feel is just as important as to know; we need to honor the emotions that come into play through cognitive dissonance.

3   *Discovering rhythm and scale.* Through engagement with soil, we can tune into the natural rhythms and cycles of the earth, moon, and sun. Living soil also teaches us something about appropriate and functional scale. Engaged with

soil in learning gardens, we see in plain sight that *maximum* size is very different from *optimum* size. Modern education is set to the lifeless ringing of bells, the tick of clocks, and rigidity of curriculum aligned with linear development. In terms of scale, in schools as in other areas of modern Western society, bigger is thought to be always better. Through bringing living soil to the center of education, we begin to transition learning toward natural rhythms and a more functional scale.

4   *Valuing biocultural diversity.* Living soil supports both biological diversity and cultural diversity, and in turn biological and cultural diversity tend to support soil. We can explore the interconnections between biological and cultural relations as manifested through interactions with soil and food in learning gardens. One single teaspoon of living soil carries a vast quantity of highly organized biological and cultural information. The immense diversity of fungi, bacteria, and insects in soil ensures that decomposition and soil creation continues. Diversity strengthens ecosystems and social systems through building complex networks of interdependence. Homogenization weakens ecosystems and social systems. Typically in schools, diversity is honored only in terms of racial, gender, or cultural diversity. Observing the vital connection between biological and cultural diversity evident in soil, we seek to broaden the conversation.

5   *Embracing practical experience.* Gardens encourage children to go outside of the classroom and put knowledge into practice. Modern educational systems tend to divorce knowledge from lived experience and application. Much learning is expected to take place as the body sits still for seven to eight hours per day. The claim made by many students at all levels that school is "irrelevant" to life rings true considering ways in which information and knowledge are generally abstracted from application. Experience deepens learning through creating a back and forth movement between old and new ways of knowing. Living soil in learning gardens must be experienced to be known. By moving soil to the center of education as a root metaphor, we assert the irreducible value of experience as part of the learning process.

6   *Nuturing interconnectedness.* Soil is a dynamic ecosystem of endless interconnections. Nested systems and patterns within patterns abound. We can become aware of the network as a defining feature of life's systems in contact with soil. Ignorance of the interrelatedness of all living things can lead to myopic action and undermines meaningful education. At the school level, subject matter is too often abstracted from a broader picture of whole-systems knowledge. Modern educational systems divide knowledge into specializations, such that experts are produced knowing little beyond their realm of specialization. Through contact with soil in learning gardens, we remind ourselves that the whole is greater than the sum of its parts.

7   *Awakening the senses.* Our senses of smell, touch, hearing, and taste are all significant ways of perceiving information and understanding the world

around us. They assist us with embodied learning, developing in us sensual awareness, often unspoken. Too often we forget to use these senses in the process of learning especially because schools, and hence learning, are mediated by the influence of technology. Electronic modes of communication highly stimulate mostly the eyes and to some degree also the ears. However, at what cost? In simply stimulating the eye through reading the text or watching a screen, and infiltrating our auditory canals with sound bytes of varying magnitudes, modern education ignores the fundamental aspect of what makes us human—our rich sensory capacities. Living soil asks of us to use more than just our eyes and ears, though these are important too. We can awaken and sharpen all of our senses through intentional engagement with life in learning gardens.

We offer this articulated ecological paradigm rooted in the soil as a regenerative alternative to the mechanistic model that now guides educational reform. These seven attributes of living soil, as seen in Figure 3.1, provide a map for education to deepen theoretical discussion about learning gardens. Manifested in learning gardens, the attributes link pedagogy and pedology. In turn, learning gardens emerge from living soil, which thus forms the foundation for both literal gardens and innovative educational practice designed with a view to bring life to schools.

Composting Activity for Building Soil: Seven Pedagogical Principles in Action

It is of utmost importance that we seek to regain a tangible understanding of soil and create favorable conditions for students to engage with and be engaged by living soil. Through our personal involvement with school learning gardens we have observed the possibilities of creating fertile grounds for sustainability education.

While food is the most palatable product of gardens, compost is the most desirable. Since long neglected soil on school grounds is often nutrient deficient or polluted, active composting makes a contribution to living soil which sustains related human and biotic communities. Unfortunately, the gardening season is out of sync with the academic school calendar; just as students are arriving for classes, the rich abundance of the summer fades to withering stalks and muddy fields. While this can be an obstacle for educators seeking to integrate gardens into their practice, it presents an opportunity for compost-making, which sets in motion a long-term investment in living soil.

Imagine a fall day in the garden, where 20 sixth-grade students are busily harvesting ripe produce in small groups led by teachers and community volunteers. There are a number of work stations, including picking pumpkins, mulching fruit trees, and building a fall compost heap. Not many students are drawn to the compost heap, perhaps because it is "dirty," but eventually two students—Santiago and Katie—agree reluctantly to help Rick, a community volunteer, gather different types of biomass for the pile. The trio retrieves a wheelbarrow and begins to gather fall leaves from the small orchard.

Katie notices that underneath the moist leaves there are many organisms such as millipedes and sow bugs. At first, she is nervous to touch them, but soon overcomes her fear. Rick explains that moist leaves are a natural habitat for decomposers, and that the compost heap that they are building is an ideal home for these organisms to flourish. Santiago gets excited managing to steady the wheelbarrow when it is filled with donated rabbit manure. Though he is first disgusted by the mixture of straw and manure, he soon finds pride in being strong enough in body and spirit to fill and pilot the wheelbarrow.

Back at the compost heap, Katie and Santiago work together to cautiously combine their gathered biomass in a careful formula presented by Rick. Other students notice their project and inquire about what they are doing. Katie explains that the decomposers are just like humans, they need food, water, and air to live. Santiago shows his friends how to add layers of leaves covered with layers of manure. Soon many students are gathered around the growing compost heap, helping to water it and keep it within the bounds of the wooden bin.

Some students are brave enough to reach a gloved hand into a nearby compost heap that is more established; they notice it is hot. Removing their gloves, they remark with surprise that the compost does not smell and that they cannot recognize any leaves or straw in the maturing heap. They wonder aloud how long it would take to transform the rough pile of leaves, sticks, and straw into one that looks, feels, and smells "just like dirt." The garden period ends and the students and their teacher return to the school building for the rest of their day. But the lesson does not end there. At snack time, Carlos, a particularly observant student, announces to the class that their apple cores can be added to the compost heap; the class community finds a way to collect the cores. Compost now enters classroom walls as students and teacher reconnect with the core of life: living soil. Decomposition becomes as relevant as Composition. In Figure 3.2 we present a lesson sketch (adapted from Dardis, Parajuli, & Williams, 2008).

Compost-making teaches many lessons such as: change over time, cycles, decomposition, life from death, the role of micro-organisms in sustaining life, and food webs. The traditional meaning of the term "harvest" is turned on its head as students first *harvest* food waste and garden debris with which to build a compost pile, then months later *harvest* rich soil and earthworms from the bottom of the compost bin. This puzzles students and draws them into the cycles of life: "bugs" become invertebrate partners in helping to break down biomass into a form usable by plants; and compost serves as an intergenerational gift to future students and the school grounds themselves. Plus, they grow seeds in this compost-turned-soil: the miracle of life presents further bounties. Students learn

---

Lesson: "Living compost: What is it and how do we make more?"

Description
This lesson introduces students to compost and the biological processes behind it. Students make a simple compost pile and watch as it changes over the next few months. They also closely examine the critters that make compost their home.

Educational goals/skills
1. Learn what compost is and its role in the garden.
2. Learn how to identify common compost and soil organisms to appreciate them.

Activity station: Making a compost pile
1. Introduce the cycle of life and the concept of decomposition. Explain that by building a compost pile; we build a home for decomposers.
2. Ask if anyone can describe what a decomposer is or what it does.
3. Introduce the "FBI": fungus, bacteria, and insects. These are decomposers that will break down the compost pile.
4. Have participants give examples of biodegradable materials that they might throw away at home or at school (banana peel, dried leaves).
5. Ask participants to describe possible reasons to compost.
6. Introduce the "BIG FOUR": browns (e.g. leaves, straw), greens (e.g. grass clippings, food waste), air, and water.
7. Explain procedures: (1) chop materials into 6 inches or less; (2) mix browns and greens; (3) maintain moisture equal to a wrung-out sponge.
8. Have the group collect brown and green materials in separate piles.
9. Assign students various tasks such as chopping, layering browns and greens, mixing, and watering the pile.
10. Review basis of composting once pile is built.

FIGURE 3.2 Compost Lesson Plan

one positive model of environmental regeneration. *Via* composting, life's lessons simultaneously surface and find roots in the learning gardens. In Figure 3.3, a student finds life wriggling in the compost.

It is in this composting activity, where students learn to build soil, that we find the manifestation of all of the seven principles of pedagogy.

1   *Cultivating a sense of place.* Composting is an activity that embodies the spirit of living in a place. Just like a tree, which contributes to its own environment through dropping leaves which then become mulch, students engage in contributing to their local environment through collecting decaying plant materials harvested from the gardens and transforming them into life-giving fertile soil. Students involved in creating a compost heap make an intergenerational gift to soil and to future classes who will garden in the same location.

FIGURE 3.3 Wriggling life in the compost

2   *Fostering curiosity and wonder.* Soil-building stimulates curiosity, wonder and, critical questioning about the broader community context of learning in relation to life. As a seventh grade student reflects: "It is strange that people can take pride in large lawns and waste their land with simply growing and cutting grass. If we plant gardens instead, and can also grow food, we can bring wildlife and at the same time eat healthy fresh food. I am worried that bees are dying in our region; how will our flowers get pollinated? How will we have fruits?" (from Williams, 2008). Moreover, compost presents an example of the profound mystery of transformation from life to death to life.

3   *Discovering rhythm and scale.* Creating a compost heap closes the nutrient loop in the garden and reconnects activities with a cyclical sense of time. Life becomes death becomes life and so on. Garden activities do not simply stop at the first frost; rather, new activities such as building the compost heap begin as plant growth ends. Linear growth is not the only positive occurrence within the bounds of the gardens—decay is also valued. Moreover, scale takes on deeper meaning as students endeavor to build compost heaps, discovering the optimal scale for maintaining decomposition heat, they learn in bodily ways that bigger is not always better.

4   *Valuing biocultural diversity.* When students build compost heaps, they combine different materials to create a fertile soil, learning that homogeneity is biologically ineffective. Gathering diverse materials, they learn to value difference. In contrast to the metaphor of the melting pot, which reduces differences to a lowest common denominator, the compost heap combines differences to create a whole that is greater than the sum of its parts. Diversity

strengthens living systems by imparting a quality that is not possessed by any of the individual elements alone. We can broaden the educational conversation on diversity by valuing contributions of difference rather than pretending that difference does not exist.

5   *Embracing practical experience.* Compost heap construction is a task that cannot be well understood through verbal or written instruction alone. One really needs to fully engage with the physical process of collecting and combining biomass to understand how it works. Students gain skill in tandem with scientific knowledge building compost.

6   *Nurturing interconnectedness.* Noticing ways in which diverse micro-organisms interact within the compost heap stimulates interest in the secret lives of decomposers. Digging in garden beds, students proudly display earthworms they have found, knowing that their presence represents a first step in the decomposition process.

7   *Awakening the senses.* Monitoring the compost heap, students are introduced to a number of ways to evaluate the progress of decompostion. Among these ways of knowing are smelling, touching, looking, and measuring with thermometers. There are multiple indicators that can be triangulated to ensure the compost pile is decomposing properly; temperature is one of many indicators, and is not the only one. A strong smell of ammonia, for example, demonstrates an excess of nitrogen in the pile. Looking at temperature readings alone, this anomaly can easily be missed, as excess nitrogen often produces a temperature reading that appears as normal in the absence of other ways of confirming the assessment. Similarly, touching tells us something about the moisture content of the pile, which informs what type of action to take in response to below normal temperature readings. All senses must be awakened to monitor compost.

The living soil of school gardens awakens endless learning. The fertility embodied in the topsoil is the intergenerational gift of past human and biotic stewardship. Our present wealth is quite plainly predicated on this inheritance, and the life span of civilization dependent on continued responsible stewardship. Students of all ages learn that the very topsoil we are currently working with today carries forward some organic matter from 500 years ago.

The enterprise of establishing and maintaining health—both in literal agricultural and in metaphoric educational soil—thus involves an untold number of minute interactions among a teeming diversity of life forms. Education *as* sustainability, then, envisions education that is no longer conceived of as something that someone *does* to another person but is recognized as a lifelong process involving holistic experience including embodied participation and interaction with all members of the human/biotic community. When sustainability education is understood through the metaphor of living soil as a state of dynamic relationship, teaching and learning become co-occurring processes.

As a new root metaphor, living soil suggests these radically different visions of education and sustainability, possibilities that are obscured by continued reliance upon the dominant mechanistic metaphors of the modern Western paradigm. Engagement with the diversity of pedons in learning gardens may well serve as fertile ground in bringing life back to the center of the educational enterprise.

In Part II, "Learning Gardens Principles Linking Pedagogy and Pedology," we explore each of the seven principles as we build a theoretical framework and share the vibrant practices and stories of learning gardens rooted in the soils of schools. Life's lessons are captured in the regenerative voices and examples of children and youth involved with their teachers, school leaders, parents, and community members.

PART II

Learning Gardens Principles Linking Pedagogy and Pedology

4

CULTIVATING A SENSE OF PLACE

No place is to be learned like a textbook or a course in a school, and then turned away from forever on the assumption that one's knowledge is complete. What is to be known about it is without limit, and it is endlessly changing. Knowing it is therefore like breathing: it can happen, it stays real, only on the condition that it continues to happen.

(Berry, 1991, p. 75)

It's like I'm a member. I'm home. I'm safe. I'm comfortable.

(Student, Learning Gardens)

In this chapter we ground learning gardens in local school and community contexts, while simultaneously exploring the challenges for education in an era of globalization. Living soil can be a guide in framing place within an ecological context, as soil is composed in place over time and plays a significant role in creation of place, through influencing establishment of plants, land-use patterns, and cultures. Significantly, soil and culture exist in reciprocal relationship: culture contributes as much to soil life, conservation, or degradation as soil conditions contribute to cultural life and diversity of places.

Living Soil and Local Place

Living soil nurtures a vibrant biotic community endemic to place and supports locally adapted plants and food crops. The connection between food and place is often overshadowed in a global food economy, but many of us still find pleasure in food harvested from local soil and even more so when it is grown personally. Something special is carried forward in each bite of food derived from local soil—we are actually able to taste our place and connection to the biotic community.

The living soil of school learning gardens is necessarily local and bound to the climatic and cultural constitution of place. It is in the terrestrial living soil of learning gardens that the proverbial "seeds of change" germinate, grow, and thrive. Thus soil is a relevant medium for thinking and learning about place in both the literal and the metaphoric sense, as discussed in earlier chapters. Soil provides a unique multifaceted and delicious venue for developing place consciousness. School grounds can be that setting. Food and soil draw our bodily and mental attention at once to the local and ecological relations embedded in each calorie or clod. Producing food for humans, food for pollinators and wildlife, and also food for thought, learning gardens serve as sites of creative re-imagining of possible place-based relationships in the era of globalization.

The infinite variety and uniqueness of places is generated through recursive confluences of biogeography, language, climate, and culture. Observing the variations in climate, flora and fauna, dialects, and food, is how one knows a new region has been entered, and this recognition may inspire both appreciation of difference and reflection upon one's home region. Travelers often remark upon such differences when arriving in a new area, seeking to contextualize new experience in relation to place. Variation from one community to another reflects the endless potential of human ingenuity in adapting differing conditions to meet common needs such as food, shelter, and meaning. Observing the diversity found in local responses to place can sustain the senses, stimulate awareness, conjure memories, and inspire novel ideas. The uniqueness of place is embodied in the soils of diverse school learning gardens.

Groundedness in Place

Groundedness in place refers to a reciprocal relationship in which one nurtures and is nurtured by the surrounding social and ecological environment. A tree, for example, forms such a reciprocal relationship with place: while it is rooted in specific soil, bounded by contingencies of water, air, sun, space, and so forth, a tree at once contributes to shaping its own environment through shedding water, casting shade, and dropping leaves which then become mulch. Thus the tree and its terrestrial home are intimately linked, each contributing to the life of the other. Education that is grounded in place is likewise motivated by an interest in forming reciprocal relationships with the local environments where students and their families dwell. Students may, for instance, test water quality in local watersheds, serve food to homeless people in their community, participate in farmers' markets, map endemic flora and fauna, plant native plants to restore degraded natural areas, or monitor urban air quality in the school neighborhood. These activities focus attention on the local particularities of global phenomena. In this way, "place" is informed by interconnections between local relevance in a globalized world (Smith & Gruenewald, 2008). Since globalization is impacting actual local places, for us, gardens provide one fruitful and practical location to grow and cultivate a "sense of place." Individual gardens are finely tuned local expressions of phenomena such as sun, rain, wind, air, and more, all of which are common globally. Just as in each location different species of plants will flourish in response to these common environmental factors, likewise school gardens can focus attention on locally relevant aspects of common global social and ecological factors. Hunger, for instance, which is a global phenomenon, requires localized responses which will naturally vary from place to place in the same way that each local garden responds differently to environmental factors as described above.

In subtle yet profound ways, gardens communicate conditions of place. Students, teachers, and their communities are addressing interrelated global crises associated with food and nature through directed attention toward rejuvenating local school grounds and surrounding communities. In Portland, Oregon, schools are de-paving parking lots and planting gardens; in San Francisco, schools are partnering with local food groups to host farmers' markets on school grounds; in Denver, cafeterias are promoting slow food; in Chicago, "at-risk" youth are learning valuable horticulture skills; in Boston, schools in low-income neighborhoods are creating green schoolyards that are seen to positively affect academic achievement; and in Houston and other large cities, multicultural gardens are being integrated into school curriculum.

Investigating the local soil, both literally and figuratively, can be a way to begin a slow process of unearthing hidden or forgotten community history and embracing place. Literally, soil analysis helps with determining the soil pH, "the quantity of organic matter in the soil often known as *humus*, the amounts of nitrogen, phosphates, sulfate, calcium, magnesium, potassium, and other elements that are to be found in soil. This is in addition to other trace elements such as iron, boron, calcium, magnesium, etc. Gardeners are accustomed to testing their local soil via samples and assessing the quality in relation to slope, drainage, fertilizer use, texture, and color, among other variables. Similarly, we can investigate metaphoric community soil by collecting local histories of even two previous generations, by bringing grandparents and elders to the schools and school grounds to share their stories. This is happening more and more as school gardens across the country are becoming welcome places for community members. As gardens get built at schools, elders feel valued as they share their wealth of stories and oral histories about place. Their investment in schools increases. One example is the Fairview Elementary School in Denver where teachers and the principal work collaboratively with Denver Urban Gardens and the neighborhood communities and honor elders who build garden beds and support youngsters as they learn to garden, weed, and harvest. Similarly, Farmer-in-Residence programs in a number of cities are encouraging local elders to share their skills both with educators and with the children at the schools. The often negative impressions of public schools are challenged by these elders once they begin to work shoulder to shoulder with the school community on genuinely meaningful projects in the gardens and when their skills and historical knowledge are acknowledged and valued.

The high mobility of Americans creates what David Orr (1992) characterizes as a society of "residents" rather than "dwellers":

> A resident is a temporary occupant, putting down few roots and investigating little, knowing little, and perhaps caring little for the immediate locale beyond its ability to gratify. … The inhabitant, in contrast, dwells … in an intimate, organic, and mutually nurturing relationship with a place.
>
> *(pp. 130–131)*

For a resident, a map guides location. Residing in place usually carries little commitment; by contrast, dwelling means that one will be around for a while and encourages an ethic of care. For a dweller, the "map" is not the "territory." Place-based cultures are often replete with detailed stories that explain how certain places came to be as they are now. Education grounded in place connects learning with dwelling. For example, at the University of Alaska, Oscar Kawagley and Ray Barnhardt (1999) have imbued teacher education with place-based knowledge embedded in the local Alaskan tribes. Student teachers from the university are annually invited to the Old Minto Camp where they spend three weeks in the summer immersed in cross-cultural understanding that no textbook can ever teach. The Elders of the Minto tribe provide deep insight to student teachers about the ways in which they try to navigate living in "two worlds": one that is the locally derived Native world with which the elders are intimately associated, and the other that is the externally defined world enveloping their existence. As Barnhardt (2007) explains:

> The tensions between these two worlds have been at the root of many of the problems that Indigenous peoples have endured throughout the world for several centuries as the explorers, armies, traders, missionaries and teachers have imposed their world view and ways of living onto the peoples they encountered in their quest for colonial domination. These tensions between the ecologically (and thus locally) derived ways of Indigenous peoples and the macro-systems associated with colonial economic and geo-political interests are a direct reflection of the tensions between local diversity and globalization.
>
> *(pp. 113–114)*

Students are given a common experience of immersion in a culture deeply rooted in a particular place, where they develop new levels of embodied and localized understanding of indigenous cultures. By embedding themselves in a new cultural milieu in a non-threatening and guided fashion, students can set aside their own predispositions long enough to begin to see the world through other people's eyes and through their histories and stories of place. Hitherto abstract phenomena such as globalization or acid rain can thus be connected to local experience and history. In doing so, educators appreciate the time and place they actually live in as rooted within a continuum of past, present, and future.

It is often through direct experience and investigation of the flora and fauna, the soils, the seasons, the rhythms of natural cycles, the histories, and the communities within which humans live, that we develop and begin to feel a sense of place. According to poet Gary Snyder (1990), "the small lessons, the enormous lessons, the lessons that may be crucial to the planet's persistence" are learned in interaction with place (p. 26). He urges that we intimately reacquaint ourselves with habitat. But, how does one develop place consciousness? For Snyder (1990): "to know the spirit of a place is to realize that you are a part of a part and that the whole is made of parts, each of which is whole. You start with the part you are whole in" (p. 38). Place-based relationships are founded in the depths of geographical, historical, seasonal, ecological, and cultural understandings over time. These overlapping understandings of place together contribute to whole-systems perspectives in which the hidden connections among geography, ecology, history, and culture become visible.

Ecological Perspective

Cognizant of the interconnection of all things and the irreducible context of place, environmental thinkers such as Aldo Leopold, Wes Jackson, Wendell Berry, Gustavo Esteva, and Vandana Shiva have encouraged a "land ethic." School gardens can provide a site for students to come to view their actions as inextricably linked with an elaborate web of life seething beneath their feet. They can learn to think like soil. Working together in learning gardens, children have an opportunity to develop place consciousness while practicing caring relationships for self, soil, and society. By learning about place in this manner, we connect with the ways our forbearers lived in community for generations. As Shirley Abbott (1983) tells us, "our ancestors dwell in the attics of our brains as they do in the spiraling chains of knowledge hidden in every cell of our bodies" (p. 1). Such cultural memory is similarly encoded within place, writ large upon the landscape, resulting in critical examination of history. Naturalist Tom Wessels (1999) eloquently explains how with a well-trained eye one may "read the forested landscape" as if it were a history book detailing previous land-use practices and associated changes of regional social organization.

Developing a "sense of place" tends to support an associated responsibility or stewardship of endemic resources, be they ecological, economic, or social, and to some degree may extend to broader consideration of and collaboration with those communities "upstream and downstream." Humans protect what they love, whether the loved subject is a watershed or a main street. But a deep knowing of place cannot be rushed in a short period of time; indeed, as Wendell Berry notes in reference to his Kentucky family farm, a lifetime may not be enough to fully appreciate the intricacies and subtle nuances of even a small plot of land.

In the global age, it appears ever more difficult to develop a sense of place, as the logic of global capitalism instigates endless relocations and smoothes out variations between places. Local context can be marginalized; conversely, inter-dependent links between local and global are increasingly evidenced by climate change, financial meltdown, and food contamination. In Thomas Friedman's (2005) "flat" world, there is little need to develop a sense of place because all places are perceived as equal, manifested by global corporate monoculture. Yet, in reality the so-called "flat world" is more unequal, more hierarchical, and more exploitative (Shiva, 2005). Absence of place consciousness has significant consequences for ecological systems and ties of social networks endemic to specific locations. As persuasive as global outreach and thinking globally may sound in the context of an economically and technologically networked world, it is too easy to forget the impact of global monoculture on local places—even more so when we are disconnected from any sense of place.

In an era of globalization, one practical solution for revitalizing the local is taking care of the soil directly under our feet. We are neither trapped by romantic illusion of local place nor enamored by the glorified "flatness" of Friedman's (2005) globe. Entering into respectful reciprocal relationships with *both* the local and the global, we reject the false choice between the two and encourage as an alternative "thinking ecologically" (Williams, 1995).

Think Ecologically, Act Locally

Like the "seeds of change" described above that require a terrestrial soil to germinate, so any action necessarily requires a place to occur. Thus globalization is enacted locally in communities around the world, as are acts of resistance. In terms of global interconnectedness, local actions inevitably affect and are affected by the activities of other locales that may not necessarily be geographically close. As Wendell Berry puts it, "we all live downstream." In turn, we also all live upstream from other communities. Thus focusing only on the local may obfuscate responsibility beyond the boundaries of defined communities (Williams, 1995). The effects of globalization upon local social and ecological systems are many. Small diversified farms tucked into hills and valleys and attuned to place-specific micro-climates and local palates are undersold by larger corporate farms, subsumed by agribusiness and forced into bankruptcy. Small businesses are similarly displaced by monolithic corporate behemoths more concerned with the bottom line than meeting local needs for goods or services. Even the efficacy of the environmental movement has been diluted by the loss of place: local resource stewardship has been replaced by sending checks to large-scale exogenous advocacy groups, such as Greenpeace or Sierra Club. Globalization also affects education, as striving for "global competitiveness" overtakes development of local knowledge as a leading educational priority. Successive waves of educational reform, such as *Race to the Top* and *No Child Left Behind*, have continually streamlined teaching and learning toward narrower and narrower objectives aimed at Americans becoming competitive in global markets while forgoing examination or consideration of the various local contexts in which people actually live.

However, a number of questions surround place as a construct. In the era of globalization, what does it mean to be "local," or to have a "sense of place"? How are constructs of place or locale defined and understood within the context of an interconnected world? Over the years, a number of suggestive slogans have emerged that seek to address the dichotomy between the global and the local. The most well known of these is "think globally, act locally." This mantra was introduced during the 1990s, communicating a form of resistance to the detrimental aspects of globalization combined with reclamation of the local, carefully avoiding full rejection of the global. Ensuing debates grappled with the contradictions inherent in reconciling the global and the local in these terms. For instance, Madhu Prakash (1994) challenged the notion of "thinking globally," pointing out the human impossibility of the first half of the directive. As she writes, "while assuming the moral obligation to engage in 'global thinking,' ... the philosophy encapsulated in the second half of the slogan implicitly warns against the arrogance, the far-fetched and dangerous fantasy of acting globally, and urges us to respect the limits of local action" (Prakash, 1994, p. 50). Prakash replaces the image of global thinking with local thinking, and suggests instead that we attempt to "think locally, act locally."

But, what does it mean to either "think globally" or "think locally"? Is thinking globally about tuning our minds to abstract phenomena without recognition of where we are or does it mean that we should be concerned with everything everywhere? This is humanly impossible to do. On the other hand, what does it mean to "think locally?" Is it about considering only our own interests at the expense of neighbors near and far? What would be the implication of this line of thinking in view of the social stratification we already see occurring? Clearly, major questions arise with respect to either thinking globally or thinking locally. Yet, acknowledging the power of the revised mantra—think locally, act locally—we ask, how can prioritizing the "local" avoid the dangers of parochialism? We believe that action must be limited to local scale. In some ways the notion of thinking and acting locally is simplistic. It overlooks the historical fact that all human communities—including oft romanticized indigenous communities—exist in relationship with neighboring groups. Furthermore, valuing the local alone carries forward a naïve American mythology of self-sufficiency and individualistic autonomy: the go-it-alone mentality. Finally, valuing the local alone forgets that idealized "place" does not necessarily meet the needs of all people; for example, the local can be homogenous, exclusive, abusive, and dominating (especially over women). It is a very fine line that separates high-minded "localism" from narrow-minded parochialism. Questions of appropriate use stem from discussion of contrasting visions of place, and often there is no simple right answer. What is evident, though, is that place consists of an infinite array of complex interconnections and relationships which can too easily be severed by ill-informed or un-judicious action. At issue here then is more than a spatial consideration of place as a finite location. Also needed is a comprehension of the timescale within which place co-evolves with the human and biotic community. Thus knowing a place might be classified as a form of "slow knowledge" (Orr, 1994) developed in relationship over time.

Local actions must be based on thinking that goes beyond the local. According to Williams (1995):

> Since we are affected by what happens upstream, we must not only be aware but also willing to act based on the knowledge of what happens upstream. Similarly, we are morally obliged to recognize that our actions will likely affect others who live downstream from us. This kind of recognition, comprehension, and thinking about our interconnectedness is extremely crucial as we *locally* begin to address the ecological problems that we confront. However, if we were to indulge solely in local thinking, then the very sort of individualism that has brought us into our present ecological imperilment would be further intensified.
>
> *(p. 53)*

For us, "think ecologically, act locally" is a more appropriate and nuanced moral injunction; in the learning gardens we have found a synthesis of the local and global vis-à-vis a whole-systems ecological perspective.

Pedagogical Implications

The above discussion urges deeper reflection and clarification on the meaning of place in the era of globalization. "When the narrative of globalization becomes effectively linked with the narrative of social justice and equity, globalization becomes increasingly difficult to challenge," according to Gregory Smith and David Gruenwald (2008, pp. xv). Recognizing this, they call for a "critical pedagogy of place" which represents a fusion of critical pedagogy and place-based education, drawing from a wide array of disciplinary sources to ignite children's desire to learn. As defined within this practice, place includes not only the physical landscape and ecology where students live, but also the cultural and socioeconomic communities that live within that landscape. The critical strand of pedagogy of place serves to bring local practices into the light of thoughtful question, while the place-based element serves to root such inquiry in the human and biotic communities in which schools physically exist. Thus students might critically investigate intangible macro issues associated with globalization through examining local manifestations in their community and their schools. For instance, middle and high school artists in Steelville, Missouri, created over 100 pottery bowls for the annual Empty Bowls dinner in the community, a way to make their communities aware of hunger and the need to feed the hungry. The Empty Bowls concept has deep messages. In exchange for a simple meal of soup and bread, guests are invited to donate money toward organizations whose mission is to address food insecurity and end hunger. The guests keep the hand-crafted bowl as a reminder of the empty bowls in their community and elsewhere. In San Francisco, Aptos Middle School students not only grow food but they also box the produce and offer it to low-income communities at local farmers' markets or take it to food banks and hunger shelters. Boston public high schools students cook and serve food to hungry Bostonians at the soup kitchen. Through such activities across the country, students dig deeper into questions of waste and preservation of food and, more importantly, they are exposed to the hitherto-invisible members of society. With their teachers, they explore what it means for a certain segment of the population to live on the streets, homeless and hungry, when, within blocks of their school, there are million-dollar mansions where food is not an issue of survival. Whereas *place* is more tangible, discussions related to solutions can take students either walkable blocks or over miles across cities to State assemblies and voting booths in search of justice. For Julian Agyeman (2005), justice and equity must be at the center of sustainability and must include egalitarianism. In decentralized education, hunger and larger social issues of inequity can too easily be theorized and uprooted from the actual communities in which education exists.

Within modern educational systems, knowledge is too often de-contextualized from experience and disaggregated from an ecology of ideas in a number of ways. For example: knowledge is increasingly divided into subspecialties; modern schools are bounded entities that are impermeable and separated from the rest of society; and teaching and learning are spatially and conceptually removed from the local human and biotic communities in which schools physically exist. The standardization of curricula, imported or exported as a stock commodity, emphasizes the industrial quality of schools, and marginalizes the critical importance of context (Esteva & Prakash, 1998; Gruenewald, 2008; Smith & Williams, 1999). As Orr (1992) explains, place has not been recognized in education; in fact, "place has no particular standing in contemporary education" (p. 126). In an age where big is considered to be better, where consolidation of organizations, institutions, and corporations is the norm, where large-scale devours anything small-scale with the mantra of "economies of scale" justifying such moves, it is no surprise that even education succumbs to the distant "global" scope of things. "Place is nebulous to educators because to a great extent we are a deplaced people for whom our immediate places are no longer sources of food, water, livelihood, energy, materials, friends, recreation, or sacred inspiration" (Orr, 1992, p. 126). As a result, the loss of context contributes to an abstraction of knowledge. Context is one of the keys to transformative learning identified by radical Brazilian educator Paulo Freire (1970). As he explains in *Pedagogy of the Oppressed*, students' ability to read and make sense of words is aided by unpacking and animating their social meaning in terms of local context:

> There is no such thing as a neutral education process. Education either functions as an instrument which is used to facilitate the integration of generations into the logic of the present system and bring about conformity to it, or it becomes the "practice of freedom," the means by which men and women deal critically with reality and discover how to participate in the transformation of their world.
>
> *(Freire, 1970, p. 34)*

While the present discourse of reform in education is about competition and preparing students for a global marketplace, there is little consideration of the consequences of such an orientation to the physical places where people dwell. Instead of the distant goals of preparing "world-class" citizens increasingly through web-based programs, in our ongoing work, we have found learning gardens as one viable alternative that helps to root children and youth in the local soils of life under their feet, encouraging "soil-based" citizenship. Based on Freire's insight, place-based and critical educators suggest deepening learning through reading the *text* in concert with the *context*, and reading the *word* in parallel with the *world*. We add that when people are rooted in place and teaching and learning are positioned in relationship to relevant local context, education will be most effective. Living soil and learning gardens can be such a context located directly on the school grounds, and can be instrumental in bringing life to schools. Place requires a pedagogy and curriculum that integrates intellectual undertakings with experience and the art of living well where we are.

Place-Based Education and a Sense of Place

In response to the pervasive power of globalization upon social, ecological, economic, and educational systems, place-based education has emerged as a pragmatic alternative to chasing a dream of global knowledge. Place-based education helps students understand complex phenomena and subject matter mediated through connection to their local communities and natural world of which they are a part. David Sobel (2004) explains: "The history, folk culture, social problems, economic, and aesthetics of the community and its environment are all on the [school] agenda" (p. 9). Indeed, much can be gained from the personal relationships and emotional attachment to place that is created through experience. For Sobel (2004) place-based education uses "the local community and environment as a starting point to teach the concepts of language arts, mathematics, social studies, science, and other subjects across the curriculum" (p. 7). According to Smith (2002):

> The primary value of place based education lies in the way that it serves to strengthen children's connections to others and to the regions in which they live. ... By reconnecting rather than separating children from the world, place based education serves both individuals and communities, helping individuals to experience the value they hold for others and allowing communities to benefit from the commitment and contributions of their members.
>
> *(p. 594)*

Unlike environmental education programs, which often occur outside of the school day and are seen as extra-curricular activities, place-based education occurs within the classroom and the surrounding community including the school grounds. Reflecting on the value of place-based education, Louise Chawla (2006) describes how, beyond the four walls of a school, a person "encounters a dynamic, dense, multisensory flow of diversely structured information" (p. 67). This vibrant "living textbook" of information frames teaching and learning for place-based education, and re-animates the world. Place-based education is built upon the experiential pedagogies of Dewey and Freire.

Examples from Learning Gardens

School garden projects are one way to bring place-based learning rooted in living soil directly into educational processes as the basis of an ecologically sustainable and locally relevant interdisciplinary curriculum. The *Common Roots* project (Kiefer & Kemple, 1998), in northeastern Vermont, addresses interrelated problems of community food security and land stewardship through encouraging intergenerational and multicultural learning with the garden as the focal point of all investigation. A central question that guides the project is simply "what has happened on this piece of land?" From here, further questions emerge such as "where are we?" "who are we?" and "where are we going?" Orienting learning to address these questions starkly contrasts the more standard individualist outlook of modern Western culture and education, and focuses attention toward community rooted in place. These questions support interdisciplinary teaching and learning woven throughout the school garden.

No two school learning gardens are the same. Variables such as differing curricular goals, water resources, cultural characteristics, weather patterns, and geographic latitudinal and longitudinal coordinates, inevitably make gardens place-specific and responsive to local needs. School gardens around the country demonstrate countless variations in application. This begins with the observation of the obvious: different types of gardens are grown in Texas, New York, or Chicago as schools are situated within unique geographical and biocultural regions requiring appropriate garden design. In our experience, these garden-based programs help students learn math, literacy, life science, earth science, and more, with strong partnerships that support the classroom teacher at all of the schools with garden programs. These school gardens, grounded in the local, are meant to enhance the health and well-being of both children and adults in the community; promote wellness through increased vegetable and fruit consumption; and simultaneously strengthen social connections especially with elders in their school communities.

A Storyline Unit: Celebrating Place

The day after Labor Day, in September, the new school year brings anticipation for first and second graders in blended classes at Sunnyside Environmental School in Portland, Oregon. The school grounds are framed by late blossoming flowers with opposing stands of ripe raspberry canes creating a welcoming entryway at the front of the building where bright yellow sunflowers with stalks that rise 6 to 10 feet greet them at the door. The soils of the gardens are inviting, with summer produce and flowers presenting life in its many mysteries and beauties: bees and lady bugs, swallows and hummingbirds, butterflies and beetles are but a few visible indicators that life continued in the school gardens even when most students were on vacation during the summer. Five teachers start the academic year with the first and second graders using the joint storyline unit, one that culminates in participating in the school Harvest Fair at the end of October. A storyline is a form of teaching and learning that is open-ended and actively engages students in producing their own visual texts, often in the form of a class mural that evolves daily. Students use all their senses to explore the environment and to express their ideas as they collectively imagine and discover even as the teacher structures the activities through prompts, reading stories, and episodes arising from the main themes.

The teachers' plan comes to life and varies from class to class. Since *place* is a significant part of the curriculum, the unit focuses on students' understanding of their own place. They begin by collectively imagining what kind of a city they would like to build and live in. Place-making develops new meaning. The students in each of the classes face a blank wall of bulletin board covered with plain paper. With their teachers they read *Curious Garden* (Brown, 2009). The story is about a little boy, Liam, who lives in a dull dreary city that has no trees or greenery and is out exploring when all of a sudden he notices plants peeking out of the ground near abandoned railroad tracks. Since the plants appear to be dying, Liam waters them only to discover that the plants do not stay in one place—they spread across the railway line as they peek through concrete, steel, bricks and gravel. Nature wants to explore the city, the story implies, and as a result, others in the city get involved, tending and caring for the plants and gardens that soon grow and begin to thrive. The important message of the book is: How would the city change if people cooperated with nature? A garden's curiosity is inextricably linked with a child's curiosity and willingness to respond to nature as a small boy helps to transform the dull, lifeless city into a luscious green one.

Modeled after the character in the book, the students discuss and get ready to transform the blank bulletin board into their city as they work on projects in groups. The city they build is renamed; each class creates its wall depicting the transformation and each new city has places for growing food and gardens. One class has changed the name "Gray" City to "Green Sunnyside" City as described in the quotes below and seen in Figure 4.1. Liam's *Curious Garden* has inspired them:

> Gray city was grey and gloomy. No colors, no plants, no trees, no birds, no animals.

> People from green, happy cities moved in. They planted gardens with tomatoes, carrots, blueberries, and other fruits and vegetables. Gray city was starting to change.

FIGURE 4.1 Green Sunnyside City comes alive on the children's wall of imagination

Flowers began to pop up all over the city. They appeared everywhere. There were poppies, asters, cosmos, and sunflowers.

The children planted trees and bushes and the animals came! There were birds flying around the city.

The Gray City was grey no more. They changed its name to Green Sunnyside City.

Simultaneously, students learn about seasonal variation over the period of two months as the autumn colors begin to set in. They regularly walk the neighborhood to observe changes in air, flora, and fauna; each time, this routine brings surprises. As part of the curriculum, students observe neighborhood trees changing color. They learn the names of different shrubs and trees, they feel the textures of leaves, they take deep breaths to smell the autumn air, they put on their artist hats to closely observe leaf shapes, sizes, and colors, and they look for signs of the changing season all around them. Autumn becomes a noticeably different experience than summer. They begin to get in tune with seasons as expressed in place.

On school grounds, they dirty their hands in the living soil of their gardens as they make bodily connections to place by digging potatoes, harvesting squash, gathering greens, or sorting sunflower seeds. Their appreciation for labor continues beyond Labor Day: working the soil, building compost, growing food, harvesting, and cooking, are all integral parts of the curriculum. At this time of year they are mostly harvesting ripe produce and planting fall crops such as fava beans. The students learn science, math, and language arts as they investigate the Three Sisters Garden: bean, squash/pumpkin, and corn, which were seeded by the previous grades prior to the summer vacation, are now growing. They visit a two-acre local farm. Back on school grounds, they learn about the parts of plants by actually planting and growing winter crops such as garlic, spinach, broccoli, and more.

Class trips to local farms expose students to the richness of nearby land and the toughness of local farmers whose strength lies in their healthy food and healthy bodies. Soil and earth come alive in ways that only tiny seeds can know as they are transformed to corn stalks, bean tendrils, and pumpkin trails. The sights and sounds, the textures and tastes, linger on for days in the classrooms as the city begins to take shape within the creative walls of the school. These walls lose their metaphoric blockades and hardness. They are permeable to life's influences as children learn to read, write, and do arithmetic in meaningful ways connected to place. Students undertake group projects, studying not only corns, beans, and pumpkins but also other plants they harvest from the gardens such as potatoes and carrots. In the rest of the grades, a similar process ensures that learning thrives as students inquire about their place through connections with local foods that they grow and/or harvest. The storyline projects culminate with a community-wide

Harvest Fair that draws hundreds of parents and the community to the classrooms and the cafeteria, the auditorium and the school grounds. Food and gardens are integrated in nature journals, group projects, and garden work right where the school is placed. Learning is rooted in soil, community, and place.

One group of students studies corn with soil, learning gardens, and life as an integral part of the learning environment. As the detailed art work in Figure 4.3 indicates, students are astute observers of the plant. They study the life cycle of a corn by actually having planted corn seeds, checking their own height next to a grown plant in a field, and estimating and counting seeds on a cob. They taste corn dishes. They even take an inventory and do simple collating of responses as they ask their peers: How do you like to eat corn? They find that their classmates' preferences are as follows: (a) fresh on the cob (three responses); (b) corn bread (six responses); or (c) tortilla chips (11 responses). They note that corn grows fast: from seed to harvest, in 80 days. They understand that each corn silk is tubular and is connected to a kernel. They also learn that 75% of all groceries have some corn in them and that much of the corn grown is used to feed animals. They become aware of the subtle differences in corn, as seen in the drawings in Figure 4.2: sweetcorn, popcorn, flintcorn, and dentcorn. Finally, they have a quiz for their

FIGURE 4.2 The husk, silk, seed, and cob of corn

classmates: Where did corn come from? (Mexco, 10,000 years old.) Is corn a grass? (Yes.) How tall can a corn grow? (Up to 20 feet.) They proudly display their understanding with drawings and attractive poster boards.

Similar projects are developed in groups for pumpkins, potatoes, beans, and apples. All of these are explained by the students to parents and visitors from the community at the Harvest Fair. Real vegetables, flowers, seeds, and plants are actually touched, felt, and observed feeding children's curiosity even further as they work on group projects. The classroom itself is alive as life is brought into the center of the learning environment.

Even as they are grounded in place, students learn about the intricate web of global connections impacting our food chain. The global links with the local are evident as fifth graders research harvest rituals around the globe and present a map that shows these (Figure 4.3). Their project shows such diverse rituals as Egyptian farmers harvesting corn and the dedication ritual for their god, Min, to the Slovakian ritual of offering a shock of wheat to the animals and mice in the field, so that they would not go the barn in search of food.

The projects described above develop place consciousness in a number of ways. Through transforming their classroom from Gray City to Green Sunnyside City, students make place meaningful. They assert shared values that frame the room that will be their learning space for the school year, an act which requires

FIGURE 4.3 Rituals of harvest around the globe

commitment to place. At the same time, in preparation for the Harvest Fair, they begin to contextualize locally grown crops and cultural traditions within the framework of a globalized food economy and diverse world. Visits to local farmers emphasize a sense of rootedness in community while learning about the origins of foods such as corn remind students of vast interconnections that span geographies, ecologies, and cultures. Beyond food, school gardens take care of local needs and issues such as rainwater harvesting, wildlife attraction, and native plant restoration. All around the country in large and small school districts, educators are bringing place into curriculum as recommended by Sobel, Smith, Gruenewald, Orr, and others, by creating gardens focused on relevant local issues.

Closing the Loop: Learning Gardens, Living Soil, and Sustainability Education

Groundedness in place can contribute to students' understanding of where they are, and thus provide direction regarding the associated question of identity in relation to place and community. As Wendell Berry has noted, if we do not know where we are, it is hard to know who we are. Developing a sense of place fosters commitment to locale, encouraging reciprocal relationships with people and land. When we know where we are and feel connected to human and biotic communities—including soil, our source of food—we are more likely to carry an ethic of care, and less likely to act irresponsibly and wreak social or ecological havoc in our communities (Berry, 1990). Place-based learning is foundational to sustainability education, as it clarifies the links between knowledge and its application: no longer an abstraction, place-based knowledge is tangible. Because it can be seen right here right now, it inspires active responsibility to preserve the integrity of this particular place and by extension all other places.

As educators, we do not claim that one cannot move from place, or that place-based education is only meaningful to those who make long-term commitments to place. We recognize that children and families do not necessarily have a choice in where they live or why or when they move. It is counterproductive to insist that they unconditionally make commitment to specific locations. However, place-based education can positively affect children's interest in specific places and can instill a certain sensibility to navigate any place with care. For instance, a child who learns respect for a local watershed and foodshed in one location is more likely to understand and appreciate the types of reciprocal relationships that sustain these networks in another location. Place-based education is a pedagogical approach that permits one to seek answers specific to locale, yet it generates a quality of knowledge that is transferable.

Place-based education supports students in becoming "dwellers" and in establishing reciprocal relationships with the human and biotic communities in which their schools are housed. Investigating certain plants such as corn and visiting surrounding farms, students begin to appreciate that they are a part of something bigger and unique. Harvest Fairs in particular are celebratory events that encourage students, parents, and communities to come together and invest in place. Because Harvest Fairs are celebrated globally (see Figure 4.3), they remain specific to local conditions, customs, and seasonal crops, thus illustrating a dynamic dialectic between the global and the local. In our work, we have seen students work in teams and groups raising questions and expressing curiosity regarding relationships between plants, food, and place, and have noted that academic work thus inspired can integrate disparate disciplines as diverse as arts, science, mathematics, language arts, and history, converging with relevance to place, as this chapter has explored. Engaged with content animated by local interest or tangibility, students are active learners participating in developing understandings of life rooted in place. Walking the neighborhood routinely observing seasonal changes, for example, not only creates a classroom community but critically awakens students to what is alive in their community here and now. At a time when most young people can recognize 100 corporate logos but fail to identify a single endemic tree species, becoming acquainted with the local flora and fauna—noticing that squirrels prefer oak trees to conifers, seeing that some trees hold their leaves longer into the autumn—is indeed significant academic and ecological learning, and a step toward bringing life to the center of the educational enterprise. Links between learning gardens, living soil, and sustainability education emerge as students connect with place. Balancing global information with local knowledge, they gracefully enact Gary Snyder's suggestion to "begin with the part you are whole in." Place serves to direct children's curiosity and wonder toward life's questions, as explored in the following chapter.

5

FOSTERING CURIOSITY AND WONDER

[In] our teaching and curricula it is only rarely that a child discovers how thoroughly in every quarter our knowledge is an act of imagination and interpretation. We are disposed to believe that the precinct of wonder is destined to shrink as the limits of our knowledge expand and mystery will be found only at those remote and distant frontiers of study. This is nonsense, as every wide-eyed child can demonstrate.

(Green, 1971, p. 202)

I wonder if the garden will help with any new inventions like a medicine or eco-friendly gas. I know that ... the garden produces them.

(Student, Learning Gardens)

To be curious and to wonder are foundational to learning. In bringing life to schools, curiosity and wonder play significant roles, returning an adequate sense of awe and mystery to an environment where the search for the right answer among predetermined choices in multiple choice tests stifles children's latent interest in life. In this chapter, we explore the relevance of curiosity and wonder in connecting children and youth to nature via school gardens and also discuss their interrelationships as foundational aspects of learning and teaching.

Curiosity and Wonder: Motivators for Learning

Fostering wonder for the endless beauties of our earthly home is, as Rachel Carson (1956) says, wholesome and necessary. However, wonder has received less attention in pedagogy since the concept is more elusive and viewed as not result-oriented. But, if we focus on the process of learning, and the fact that students' knowledge always evolves and is incomplete, then wonder plays a significant part in learning and needs to be encouraged. Unfortunately, being driven by the call of the bell rather than the call of inquiry, schools are not organized to allow sufficient time for students to pursue what may hold their attention at a given moment. Wonder can lead to disruption in timed and standardized learning and loss of control on the part of authority. Yet, being in a state of wonder can often serve as a catalyst, a spark, to the act of thinking and further pursuit of knowledge. When children wonder, they begin a journey of engagement and investigation of the world. In *The Life of the Mind: Thinking*, Hannah Arendt (1971) argues that wonder is crucial to rigorous thinking and doing good philosophy. The poet and the philosopher are one, according to her:

> Because philosophy arises from awe, a philosopher is bound in his way to be a lover of myths and poetic fables. Poets and philosophers are alike in being big with wonder.
>
> *(p. 141)*

Similarly, in valuing wonder, in the classic *A Sense of Wonder,* Rachel Carson (1956) argues:

> A child's world is fresh and new and beautiful, full of wonder and excitement. It is our misfortune that for most of us that clear-eyed vision, that true instinct for what is beautiful and awe-inspiring is dimmed and even lost before we reach adulthood.
>
> *(p. 56)*

In honoring children, Carson (1956) is especially eager that adults not impose their standards and understandings and thereby hurry childhood. She adds, with regret, that adulthood brings "the boredom and disenchantments of later years, the sterile preoccupation with things that are artificial, the alienation from the sources of our strength" (p. 57). The lack of certainty and the apprehension that wonder entails are, in many ways, humbling. Wonder leads to philosophical inquiry. "Children love to ask questions related to frames, i.e. to inquire into those things we normally take for granted," according to Paul Opdal (2001, p. 333). While it cannot satisfy the modern instrumental and utilitarian ends of education, wonder feeds the appetite of the child who is hungry for exploration and often in awe of nature. Carson's love for the mysteries of nature was also driven by her profound sense of loss associated with the mindless violence toward life: "I believe that the more clearly we can focus our attention on the wonders and realities of the universe about us, the less taste we will have for destruction" (Carson, 1956, p. 58). She posits that if even one adult would share a sense of wonder with an awestruck child, then it is likely that the child will be able to maintain a natural state of wonder, an openness to the magnificence of the world, for as long as life continues.

The experience of wonder "signals that the present way of looking at things [is] incomplete" (Opdal, 2001, p. 339). This alert signal sets in motion a process to evolve new mental frameworks ever better suited for understanding a complex, diverse, and changing world with an openness toward rethinking assumptions, and an aim toward student perspective development. Wonder can also motivate uninhibited and insightful observation of phenomenological patterns expressed in nature. It is the state of mind of wondering that leads to an experience of awe and sets into motion the search for responses. Wonder relates to an open state of free thinking that transcends the boundaries of existing subject matter (Opdal, 2001). And, wonder may be more broadly concerned with the philosophic dimensions of experience, asking "what is life?" Questions are the language of wonder (Commeyras, 1995). When they question, students are thinking, seeking meaning, and connecting new ideas to familiar concepts. The constant refiguring of "why" demonstrates children's ability and willingness to wonder at the very legitimacy of frames of reference adults tend to take for granted. This is exactly the type of innovative thinking most needed at this time of "great turning" (Korten, 2006; Macy, 2009) and social and ecological uncertainty.

Curiosity, which is often accompanied by wonder, is also associated with active learning and engagement since students are regularly motivated to seek answers to questions they are curious about. In driving students to seek knowledge, to search for answers to their inquiries and fascinations, curious students look for an end to cognitive dissonance. Curiosity obviously motivates learning through continual production of questions. Neurophysiological theory suggests that curiosity is a state of arousal due to complex stimuli and uncertainty which lead to exploratory behavior. Stimulus variables such as unfamiliarity, novelty, complexity, ambiguity, and incongruity may increase arousal level and induce curiosity. For exploratory behaviors, place, locale, community, and soil provide a rich milieu.

Fostering curiosity pays off in terms of meaningful learning. When students are motivated, they are in awe and ask questions; in search of answers, they form habits of learning that support engagement with the world. This is especially important in the development of learning since ecological issues and problems have no defined and precise answers and solutions. As an iterative source of learning, gardens on school grounds particularly provide opportunities for children to organically engage in search of meaning-making as questions abound in nature. As Frederick F. Schmitt and Reza Lahroodi (2008) explain: "It is commonplace that curiosity facilitates education and inquiry, and even that frequent states of curiosity are psychologically necessary for a person's regular success in learning and discovery ... Stimulating curiosity is central to education and learning" (p. 125). Curiosity refers to a state of deepened interest about a topic or subject matter. In other words, children's curiosity may drive a search for answers within a particular research interest, for example "what are characteristics of living things in this garden?"

In articulating the value of curiosity and wonder in education, it is helpful to briefly distinguish curiosity from wonder, as they each serve to develop students' intellect rather than mold it into artificially prescribed divisions of knowledge through slightly different means. Children's latent curiosity and wonder are often manifested in asking questions (Doris, 1991), which is a link between thinking and learning. Curiosity often foregrounds an element of uncertainty and ambiguity requisite for stimulating wonder. In particular, wonder "arises from surprise at or puzzlement about the object" (Schmitt & Lahroodi, 2008, p. 128). Wonder contrasts curiosity insofar as it simply seeks to delve deeper into mysteries of experience within accepted frames of reference while curiosity questions the very frames in which meaning is constructed (Opdal, 2001). On this view, wonder may be of particular importance in the case of sustainability, as this emergent field is largely about dealing with uncertainty. In the context of compounding social and ecological crises, it seems natural and appropriate for a degree of "surprise at, or puzzlement" to characterize experience. Such relentless questioning of the existing frames of reference may therefore be most helpful in seeking solutions than further exploration of problems of sustainability within familiar understandings. It is critical to encourage and nurture curiosity and wonder within modern educational systems.

Research points to curiosity and wonder as motivating factors that interest children to construct meaning about different things (Opdal, 2001; Schmitt & Lahroodi, 2008). Engaging children's latent curiosity and wonder in cultivating more-than-intellectual awareness is thus one positive approach to be explored here with hopes for improving educational systems and inculcating a love of place and of the more-than-human world. In a time increasingly defined by escalating social and ecological crises as well as by an unprecedented access to instant information, the preservation and explicit nurturance of children's latent capacity to wonder and the ability to ask the unasked and unanswerable questions are required. Unfortunately, in a race to the top for global techno-scientific competitiveness, something is lost along the way: sufficient space for engaging with the wonders of life rather than preparing for future careers. While daily it is becoming ever more apparent that more information will not solve or reverse pernicious global problems, perhaps more questions might help, if only in opening hitherto unopened doors of inquiry and reorienting conventional patterns of perception.

Ecological Perspective

Natural settings motivate endless marvel. Learning gardens provide a landscape of inquiry directly on the school grounds that transcend the search for quantifiable answers. There will always be more to be curious about or to wonder upon in the diversity of school learning gardens. Through protracted observations students may become more interested and curious with regard to the layers upon layers of relationships embedded in even a simple untended corner of school grounds. Appreciation of relationships between and among parts as a defining characteristic of all living systems—including ourselves—can be called an ecological perspective. In this light, the term "ecological" is extended to refer to relationships and networks as defining patterns of organization not necessarily bound to conventional understandings of ecosystems. An ecological perspective encourages viewing social organizations such as families, schools, and cultures in terms of their myriad relationships and is associated with a transition from valuing of answers to questions. Through applying living soil as a guiding metaphor to sustainability education, we can further distinguish between two types of curiosity: one that delves deeper in the established disciplinary boundaries and one that seeks whole systems perspective and adds breadth of perspective to depth of understanding. Education guided by living soil as a central metaphor is enlivened by the latter form of curiosity.

Gardens can provide the tangible context for continuous pursuit of curious interest. For example, one may become curious with regard to soil micro-organisms and take up study in microbiology or soil science. Curiosity in this sense is about searching for answers. This process plainly inspires active and engaged learning on the part of students, but too often narrows the inquisitive gaze to such a degree as to block from view peripherally related information or realms of inquiry. For instance, soil micro-organisms are intimately related with: agricultural use of land, production of food, post-consumer or post-production waste streams, urban planning, and industrial use of land. Hydrology, geology, geography, ecology, and many other scientific disciplines may be of related interest and can provide additional information. Human relations with soil and soil micro-organisms are addressed in studies of cultural ecology, biocultural diversity, and ecological ethnicities. Using living soil as a metaphoric guide of sustainability education encourages a wide inclusion of interrelated disciplines as part of any such curiosity in soil micro-organisms. This view positions curiosity as an invitation to breadth of perspective in addition to depth of understanding and must be somewhat differentiated from curiosity as inquiry in terms of established disciplinary boundaries.

School grounds in the city in particular serve as a context for curious and critical inquiry. For example, many schools have expansive paved parking areas that in some places are more than adequate to serve the parking needs of the school communities. In de-paving events at schools, students, teachers, and community members transform unused parking areas into exposed soil for growing learning gardens. At one local school, Vestal Elementary in Portland, where there was more than four times the required parking lot, teachers, students, and the principal worked with city commissioners to tear down the impervious asphalt and create not only school gardens but also community gardens that serve the surrounding low-income and culturally diverse communities, thereby increasing participation of parents in schools. Activities such as reclaiming asphalted school grounds offer a hopeful glimpse into potential restorative relationships between human culture and soil, and ask us to critically consider and examine the social norm of paving soil as a sign of progress. Children may first wonder how to break slabs of asphalt into smaller and smaller pieces in order to move them, perhaps later becoming curious as to why large areas on school grounds were ever paved. This critical question might lead to historical learning, or inquiry into physical and chemical science. Once the asphalt is removed, students may wonder how plants will ever grow in the seemingly lifeless soil finally exposed to light. Or, they may already have been curious when noticing persistent plants creeping through tiny cracks in the concrete. Such opportunities for fruitful inquiry for students in making connections with their place, locale, and the natural world, abound in learning gardens.

There is certainly ample room to extend the role of student curiosity and wonder against the backdrop of standards-driven curricula. Our educational system is presently modeled after industrial processes: subject matter is standardized and learning measured through tests in which there can only be one right answer among a limited number of already predetermined choices. This framework contributes to a privileging of answers and concrete knowledge and discourages questions. How can a student openly wonder in response to a query when presented with only a set number of choices, of which only one is known to be "correct"? Classrooms profess a fact-driven curriculum rather than one that permits curiosity and wonder in learning. Plainly, the simplistic clarity of the multiple choice test, while commonplace during childhood, is rarely repeated in "real life," wherein we often encounter situations with no clear right or wrong answers. Standardization and test-taking force children into the "correct" way— the "getting it right" syndrome, instead of permitting students to explore the unknown. The start of the twenty-first century is already a time of great uncertainty, especially with regard to social and ecological systems. It is appropriate if not outright necessary to actively prepare students to explore the unknown through fostering abundant curiosity and wonder.

Links between curiosity, wonder, and critical thinking are significant. Since active curiosity and wonder are associated with critical thinking and the search for meaning, sustaining and nurturing these capacities in children is not only holistic but is also in the interest of schools. As in the example of de-paving above, often innocent interest or uninhibited asking of questions can uncover intellectual fodder for serious consideration. Working with soil in learning gardens brings students into close and intimate contact with the terrestrial source of food, and sparks lines of critical inquiry. For example, in learning gardens, we have heard young children ask penetrating questions relating to topics such as global food trade, persistent use of indiscriminate life-killing chemicals in conventional agriculture and on school grounds, high proportion of corn in the modern diet, and the ubiquity of grass lawns. Many of the questions have no simple or convincing answers. For students, engaging with life's wonders poses opportunities to ask questions instead of passively accepting the taken-for-granted cultural assumptions framed by the dominant paradigm. Creation of challenging questions is a key component in the cognitive process that contributes to certain aspects of learning, such as critical thinking. Rather than accept the rigid structures of educational institutions, the wondering child is more able to remain present to the cognitive dissonance surrounding existence and give voice to that discomfort.

Pedagogical Implications

Curiosity and wonder are aptly suited to engagement with natural systems and likewise nature is a significant source of awe. A steady diminishment in children's direct access to natural areas as a result of changing cultural patterns and broad-scale degradation of nature corresponds with a narrowing of curricular freedom in the era of schools *at risk*, NCLB and *Race to the Top*. High-status technological ways of knowing (e.g. Internet) and modes of relation (e.g. social media) now popular appear to channel curiosity and wonder in creative ways, yet our assessment is that these technologies mask a predictable and mechanistic worldview (reducible to 1s and 0s) and carry forward specifically anti-ecological cultural assumptions and root metaphors in dangerous ways. All the virtual reality in the world will never be able to replace the simple wisdom of a wondering child squealing in joy with wriggly worms in hand. School gardens are positioned as prime venues for nurturing children's tendency toward questions, which can be gently directed toward critical inquiry related to pressing environmental issues.

Examples from Learning Gardens

Direct experiences of nature urge us to engage using all of our sensory faculties in making sense of complex phenomena. Stephen R. Kellert (2005) has distinguished three distinct types of nature experience including direct, indirect and symbolic. Learning gardens and living soil encourage connection through direct experience as a way of fostering curiosity and wonder. Visits to nature centers or zoos are examples of indirect nature experience. These types of engagement provoke interest, yet often serve as distraction from issues germane to place and seem to have little connection to curriculum. Books, stories, songs, and oral histories constitute symbolic experiences. Symbolic experiences of nature can be of great significance, as in the case of oral histories, mythologies, and stories. Certainly books stimulate curiosity and wonder in deep ways, and many profound marvels are presented in this form. To us, direct experience is an important route to awakening children's sense of wonder and fascination with the natural world.

Our promotion of direct experience does not discount the value of books and other secondary media or indirect experiences such as visits to zoos or nature centers in fostering curiosity and wonder; rather, the principle here is the awakening of fascination at the seemingly ordinary and mundane, the taken-for-granted aspects of life that too often are ignored in pursuit of high-status or novel areas of inquiry. There is nothing more ordinary, mundane, and taken-for-granted than soil; yet, it is fascinating. For example, in a learning garden, a student may wonder about the life of a worm freely imagining a subterranean existence. Through becoming quiet in place in what Jon Young (2001) calls "sit spots," she may become alert and attentive to the many interactions among biotic organisms, or the subtle seasonal changes manifested in a particular garden spot, and notice relationships that had up until that point been invisible. Individuals of all ages find a "sit" spot or "sacred" spot in nature in a garden in order to learn to be quiet, and to begin to listen and observe with care and attention. This can last for ten minutes or an hour. But revisiting the same spot is the key since this encourages children and youth to sit quietly at a consistent place in nature and helps them learn to pay attention, to be mindful, to observe intently, to understand seasons, to marvel, to wonder, to listen with care, and to breathe. As Gandhi reminds us, our problems stem from our inability to sit quietly in a room alone. Our minds are conditioned to wandering. Instead, by directing focus on one place, a constant, and one sense at a time, students build sensory awareness through practice. This provides them a way to connect with the natural world. The sit spot is often called the sacred spot since it creates an awareness of wilderness that the modern hectic life, including school life, does not normally permit. As a doorway to a world of quiet, in some sense, the experience makes students more contemplative. The quality of observation stimulates the seeking for answers and the rethinking of basic assumptions; fostering curiosity and wonder thus is fundamental to learning both content and context and is invaluable to sustainability education pedagogy.

Carrying the metaphor of living soil back to the tangible terrestrial inspiration presents a wealth of unanswerable questions that can animate learning. For example: How do worms breathe underground? How is *humus* formed? How does a living soil seem to almost endlessly make room for more water, micro-organisms, bacteria, and worms, and yet not bulge to the point of eruption? These questions and more, while technically answerable, invite joyous and timeless wondering in much the way as passing clouds provide endless variations of images and likenesses. Teachers are relieved of the position of a knowing expert, since a defining characteristic of soil is our collective human ignorance of its numerous processes.

Terrariums aside, we literally know very little of what goes on below our feet. While many of the questions of soil do indeed possess specific answers, they are of a quality that one may pursue further investigation even after the "answer" is known. Life is endless questions. Living soil is full of questions. Answers lead to more questions. Each unit of soil, called a pedon, is unique, defined by place, climate, hydrology, and cultural connections. A student at work with living soil in learning gardens encounters a world in which questions have no end. She may ask "How do leaves mixed with manure turn to compost?" and yet continue to wonder at this otherwise unpredictable transformation long after an "answer" is delivered. Thomas Green (1971) explains that it is the contingency of certain phenomena that allows wonder to continue long after discovering an adequate answer. As an example, he points out that though he knows *why* grass is green he yet wonders why it cannot ever be some other color.

Living soil is contingent on numerous unseen interconnections and microscopic networks of interrelation. Perhaps it is for this reason that even the most erudite books on soil science retain a level of vagueness regarding the exact formation of *humus* in soil. Even as adults, we often find a sense of perpetual wonder and awe watching seeds transform into plants, emerging gracefully as if enacting an ancient dance from the unknown depths of soil. Indeed, a sense of disbelief that small black seeds could ever turn into large white onions is a lasting awe that does not cease to amaze students and teachers alike.

School gardens can be a convenient setting for introducing children to nature and inviting them into a reciprocal relationship with the local biotic community. Thoughtfully designed gardens are human-mediated landscapes that support a multitude of opportunities for meditation and observation. Gardens also provide a tangible context for learning about natural systems such as soil and plants in a physical way, as the following student quote indicates: "I get to work with the soil and plant. Its hands-on instead of talking about it, I get to dig and get messy." The affective dimension is an important entry point for teaching and learning sustainability. Gardens, as a site for engaging children with nature, are one place for applying this insight: there are numerous opportunities within a garden for teaching children to attend with their senses. The following student quote confirms this observation: "You feel it, hear it, touch it. Instead of looking at a book, you actually work and try to plant a plant." Below, an eighth grade student poem captures the mysteries of nature and a view of knowledge inspired by curiosity and wonder:

The Geologist

Tell me geologist,
Why do you spend your life destined to study rocks?
Because to study rocks is to know the knowledge of life.
Then tell me this geologist,
Why do we walk upon these rocks if
They are the knowledge of life?
The geologist did not answer.

Then tell me geologist,
Why do we carve upon the rocks
If they are the knowledge of life?
The geologist did not answer.

Then he lifted his head and
Peered into my eyes. His face
Was sandstone, his hair stuck up at
Places and was jet black like
Igneous, his eyes were a pale
Color and had a surface like
Moons.

Then he turned away to say,
To study rocks is to know
The knowledge of life and that
Is why I study them.

Protracted and thoughtful observation naturally stands in stark contrast to the dominant modes of thoughtless action modeled at all levels of society in modern Western culture, to which children are well accustomed. Introduction to an alternative mode of participation in the world can be quite an adjustment, but one that is all the more necessary given the complex socioecological landscape today's children will be inheriting. Adults seeking to practice observation, often follow sage advice to "adopt a child's mind," entirely open to experience. As Rachel Carson (1956) laments, while most children are naturally filled with awe, it is a great misfortune that most of us lose our innate sense of wonder before we reach adulthood. Through consistent engagement with more-than-human nature in school gardens, children may have a better chance to preserve the so-called "child's mind" further into adulthood. Only living systems can offer such knowing reflection of what it means to be alive.

In the poem "The Geologist" above, we get a glimpse into the wondering minds of young adults. The student writer considers the study of rocks as a way to connect with knowledge of life. But the study described is far from the rote memorization or analysis most common to modern schools. In her dialogue with "the geologist," the active wonder of the practicing scientist is revealed through the markings upon his face in ways his voice cannot put into words.

A Unit on Decomposition

Studying a unit on *decomposition*, in a first and second grade blended classroom at Sunnyside Environmental School in Portland, Oregon, students are grouped into teams of four to closely observe and to describe the soil on their school ground.

The teacher provides students with transparent plastic cups and prior to having them explore the schoolyard she asks them where in the outdoors they would look to bring soil in their cups for further examination under a magnifying lens. Students respond that they would most likely be able to scoop up some soil in areas that are not too dry where there has been full exposure to sun, that the best soil to sample would likely be from under a shady tree, or that puddles would probably be less likely places to scoop up soil. They squeal with joy as they go outdoors as scientists to observe and play with decaying soil as they scoop up *humus*, leaves, roots, and sticks, in varying degrees of decomposition (Figure 5.1).

Some students point out that the color of soil differs from place to place in the schoolyard. Once back in the classroom with their successful gifts of soil, they pull out magnifying lenses and start examining the soil in some detail. Several kids are thrilled to find insects and tiny bugs they had not seen with their naked eyes. Subsequently, over the week, they build terrariums in the classroom and also do experiments on what exactly soil is. They learn that soil is living and their terrariums have holes to permit air in for the soil organisms to breathe. One day, to their surprise, they find a cluster of slug eggs mysteriously sitting in a clump next to mama slug in one of the terrariums. Students are full of curiosity about how many slugs would be born. They want to know more about soil and about

FIGURE 5.1 A young scientist gathers soil on school grounds

decomposition. Collectively, the class fills up a blank bulletin board with what they are learning about decomposition and soil, and about the organisms that support the process as seen in Chapter 3. Students also learn about the various layers and kinds of soil: clay, *humus*, sub-soil, parent soil. They even look at soil under a microscope and come to realize that microbes live in soil. "They are too small to see without a microscope," students indicate on their mural.

The teacher has placed an open-ended question on a blank chart: *What do you wonder about decomposition?* The students are curious about many things and have many questions of their own that they write in their books: How does the dirt stay moist? Do worms eat roly-poly? Why is the pinecone from the soil getting fungus? Are the worms helping the plant to grow? Why are the leaves turning moldy? The teacher captures students' questions collectively on a chart for all to see and discuss:

- How can worms breathe underground?
- How do worms work?
- How many kinds of decomposers are there?
- Do all different plants decompose?
- How do things sink into the ground?
- How do decomposers decompose when it snows?
- How do moles act as decomposers?
- How many parts are there in the decomposition chain?
- How do worms go underground?
- Why is the slug above ground and worms underground?
- After a year, do worms continue to decompose the soil?

These questions, generated by students rather than the teacher, show the depth of observation and connections students make as they progress through the unit. They read storybooks about decomposers and about soil. Their music teacher also teaches them dirt songs which they sing: "Dirt Made My Lunch," and "Decompositon," each by Steven van Zandt. They do experiments and observe soil over a period of five weeks. Decomposing is a cycle; *decompose* is a scientific word for "rot." They jot down their observations. Their words are captured by the teacher for all to see: worms eat plants to decompose them; when worms tunnel, it is good for the earth; mushrooms are decomposers, too; different snails and moles are decomposers; decomposed things sink into the ground and help other things grow. There are also microbes in the soil, too small to see without a microscope (Figure 5.2).

An Experiment with Germinating Beans

Having learned a unit on beans, corn, and squash, first and second grade students in a blended class become scientifically intrigued about what it takes for beans

FIGURE 5.2 Capturing soil life

to grow. They experiment with beans in the classroom by placing beans on paper towels and in plastic bags. In some instances they water the towel and make the bags air-tight. In other cases, they do not do one nor both. The germinating bags are all left in a window exposed to sunshine. The teacher captures the students' observations which she records on a chart for all to see:

- Some beans had mold.
- Some beans had two sprouts, two molds.
- Some sprouts were light gray inside the paper towel.
- Some looked like they had hair on them. There was fluffy stuff.
- They smelled bad.
- Some had no mold.

The teacher then records questions students have about the beans based on their observations:

- How can it grow in paper towel with no dirt?
- How did it turn to mush?
- Why does it smell so bad?
- How can we tell if it needs more water?

The students then collectively propose what they would change for next time; this is recorded again, by the teacher:

- Try sprouting in soil.
- Change the paper towel every day.
- Add one drop of the water a day.
- Soak the beans before putting it on towel.
- Open the bag every day for air.

These two examples—decomposition and germinating beans—clearly show that students are actively engaged hands-on and academically with the gardens, and that their innate curiosity is fed as they seek answers, raise questions, and wonder about the life-sustaining qualities of both soil and seed.

Closing the Loop: Learning Gardens, Living Soil, and Sustainability Education

Through applying living soil as a guiding metaphor for learning gardens and sustainability education pedagogy, we indicate a need for associated redesign of the educational mindscape that shapes teaching and learning. This refers to making room in the standardized curriculum for curiosity and wonder, and moreover giving value to questions themselves. As Rilke (1934) suggests, unanswerable questions can be our guide. Engaged with life, questions lead to more questions. Thus we inquire, why is it that, as students progress in school they tend to stop asking questions and forget to wonder? Or, as essayist Neil Postman (1994) asks: "Why do children enter school as a question mark and leave as a period?" Observing children gleefully digging in soil, searching for micro-organisms with magnifying lenses, and eagerly wondering more and yet more about the fascinating processes of decomposition, inspires us and demonstrates that the pedagogical value of curiosity and wonder is unbound. The endless questions buried in living soils of learning gardens are representative of the many ideas associated with a shift toward sustainability education. If we are honest with ourselves, we can see that there is no one answer but there are many paths toward sustainability. The implication of this realization is that everybody can contribute to the generation of questions. Bringing living soil into the center of the educational enterprise encourages the fostering of curiosity and wonder at every level of education: "digging deeper" is both literal and metaphoric.

6

DISCOVERING RHYTHM AND SCALE

It is the ritual connection to nature, especially to seasons, that has always identified traditional environments and that is missing in most urban settings ... If people don't evolve complex celebrations of both time and season, they cannot fully occupy an environment. They cannot find themselves nor can they pass on the truth of a place to future generations.

(Knowles, 1999)

It is February now and the school botanist will return with more seeds. My class has prepared the garden beds with rich compost and our classroom has become a seedling factory. I look forward to the day of sitting once more on a stone and drawing the beauty of a garden unfolding.

(Student, Learning Gardens)

Together, rhythm and scale describe patterns of relationships which form the basis of our ecological model for learning gardens. Natural cycles and earthly scales are generally overshadowed in modern culture by the socially constructed artificial rhythms and scales of the dominant global economy which impacts schools. Learning set to the lock-step of a clock, and scale understood in terms of "efficiencies," undergirds modern educational systems. Instead, we need different approaches at different orders of scale that are more life-affirming. In this chapter, we use rhythm and scale as constructs drawing upon soil and gardens to recalibrate education.

Rhythm and Scale: Living Soil as Entry Point

What do rhythm and scale have to do with living soil, learning gardens, or sustainability education? Soil is not normally thought of as having any particular "rhythm." And while soil is often seen as a measurable commodity, the term "scale" rarely comes to mind. It is for these reasons that we have chosen to highlight rhythm and scale, singularly and together, as key attributes of living soil that correlate with principles of sustainability education. We see literal soil work as an entry point into understanding natural rhythms as well as developing an appreciation of humane, or earthly, scale. Using hands to establish learning gardens, the value of connecting with natural life rhythms and humane scale become abundantly clear. For example, when digging a garden bed with hand tools, one is able to literally feel the movement of the earth with the movement of our bodies. We know that we are engaged in the life-giving cycle, at once nurturing and being nurtured by earth. To dig, to hoe, to sow, and to reap, are perhaps some of the oldest defining activities of humanity. Certainly the activities of the garden are part of an elaborate continuum, one that links our physical actions with those of the generations that have come before and those that will follow after. We are able to feel the interaction of physical rhythms such as breathing, and earthly rhythms such as the movement of wind and water. As we inhale, it is as if the earth is exhaling, and vice versa. Scale also takes on greater significance when working by hand, as the amount of land one is physically able to initially prepare is limited. This often turns out to be beneficial later when weeds must be attended to, as the scale is precisely fitted to the physical capabilities of the gardener.

Meaningful tactile encounters with living soil can help to recalibrate the modern sense of time and space toward a life-enriching ecological model; they also underwrite a regenerative approach to enlivening degraded social and eco-logical systems through participatory engagement. As a metaphorical construct, living soil replaces Descartes' mechanical clock and underwrites a transition from the "age of Enlightenment" to the "age of ecology" (Kumar, 2002).

Ecological Perspective

We are surrounded by rhythms of all kinds: the daily rising and setting of the sun and moon, the turning of the planet, and the changing of the seasons are common examples of natural rhythms that frame life. The relationship between different natural rhythms and various geographical and biological scales creates endless diversity of ecosystems and *niches* in time and space. Living soil is a vibrant example of such a dynamic ecosystem hosting a diversity of micro-organisms, flora, fauna, arthropods, worms, birds, and humans in various time and space niches.

Earthly seasonal rhythms ensure recursive waves of abundance, as seen in a general garden calendar: winter enjoys preserved autumn harvest; spring awakens in succulent growth; summer boils in bounty; autumn revels in harvest. In the garden or on the farm there is always work to be done and yield to be obtained. We can generally follow the arc of the sun across the sky and the changes in length of daylight to organize activity. Days repeat within season; seasons repeat within years; and "years repeat within variations too sluggish for people to notice that things are changing around them" (Knowles, 1981, p. 5). Rather than follow the contrived tempo of clocks which generally govern our lives across the globe in a standardized way, in a garden we follow a tempo of time rooted in intention, life systems, and human activity. There is a difference between *chronos* or chrono-logical, serial time, and *kairos* or event-based time. Named after the Greek god of opportunity, *kairos* conveys a living rhythm connected with cyclical time associated with the motion of the sun, stars, moon, and the earth. Gardens are one location where we rely upon and submit to kairotic time. Seasons and soil conditions dictate what is able to grow when. As Ralph Knowles (1981) describes, on farms, "chores are more a function of sunrise and first frost than of wound springs and pendulum works" (p. 4).

Historically, human societies have lived in congruence with natural cycles and developed nuanced understandings of complex ecological rhythms. The annual flooding of fields, for example, has been predicted by movement of stars, and changes in length of daylight. In view of the interrelationship among various life rhythms, we can consider the entire planet as a self-regulating organism, inhaling and exhaling in regular intervals, circulating water and nutrients in endless cycles, and moving through space in a slow but graceful spiraling arc. NASA scientist James Lovelock (2000) has termed this perspective the Gaia hypothesis, yet, this novel scientific conception of Earth as single living organism has long been understood by diverse spiritual cosmo-visions. Understanding the Earth as a single organism, whether motivated by scientific or spiritual teachings encourages a deeper admiration and enjoyment of the elegant synchronicity of natural rhythms and cycles observed in living phenomena.

Expressions in Patterns and Cycles

Understanding patterns and relationships as central features of living systems, supports shifting attention from parts to the whole (Capra, 2002). Rhythm refers to recursive tempo, beat, pulse, cadence, and cycles. We may also describe rhythm as patterns expressed across time. The patterns we see in nature are expressions of rhythm in time translated into the physical landscape. For example, sand dunes describe the general S-curve of wind pulsing in repeating sine waves. The layered composition of soil, like sedimentary rock, encodes annual cycles of growth and decay in discernible "horizons." The rings of a tree mirror the ripple pattern of still water disturbed by a stone and so describe a slowly unfolding rhythm, or pattern in time. If the rings are to be compared to ripples in water, then the genesis of the tree can be seen as the disturbance which sets the expression of the familiar pattern in motion. Scale refers to magnitude of size, scope, and relationships of spatial dimension. Orders of scale describe subsequent articula-tions of repeating patterns, such as the branching form observed in trees or

FIGURE 6.1 Repeating spiral pattern in a cabbage

river drainages. Cutting a cabbage crosswise, we see a general spiral pattern that is repeatedly expressed in successive orders of scale at each node of the general form, as seen in Figure 6.1.

These sorts of patterns are also seen in other plants growing in school gardens. For example, the seeds of a pumpkin are aligned with the outer grooves on its skin. When children learn about this pattern, they become skilled at estimating the number of seeds through a dynamic combination of pattern recognition, geometry, and mathematics. The abstract concepts of mathematics are grounded in the relevance of physical observation and real living phenomena such as pumpkins. Another implication of an understanding of scale and order in living systems is that optimal scale is not the same as maximum physical size. Biological necessities such as heating and cooling, efficient use of nutrients and removal of waste, and maintenance of external protection (e.g. skin/hair) encourage living organisms to determine and adopt optimal rather than maximal size. Optimal size is ecologically appropriate and in balanced relationship with a number of factors such as the surrounding environment, the nuances of the season, the varying degrees of rain, soil, sun, temperature, and more. In ecological and biological terms, bigger is not always better. Scale matters.

Patterns connect rhythm and scale. Customarily we think of patterns in terms of either space or time. Examples such as spirals and waves are more easily perceived in space, whereas the beat of a song or tick-tock of a clock is recognized in time. This division is artificial and reflects more the differentiation of ways of knowing the world (i.e., eyes and ears) rather than distinct phenomena. Patterns are elaborated through time in the form of rhythmic (regular) occurrences and manifested in various orders of scale. Often the repeating tempo is slower or faster than human minds are accustomed to perceiving, which accounts for a general overlooking of the rhythmic component of patterned systems.

Living systems such as landscapes and organisms bring rhythm and scale together in applying patterns to form. Birds in flight, for example, form a chevron pattern that serves to decrease wind resistance for the group as a whole. While the lead bird bears the brunt of the wind force, the movement of air is diverted around the birds following in the chevron pattern. The flock routinely changes position to distribute the stress evenly and to optimize the group's flight. The chevron pattern of birds in flight demonstrates optimal scale, both in the number of birds in one chevron and in the distance between birds. Following seasonal changes, the flock also demonstrates rhythmic timing in its migratory movement.

Natural rhythms such as the rise and set of the sun and moon can be classified broadly as "life cycles" in that they animate and give rise to life in its myriad forms. We are all familiar with these patterns due to their daily occurrence. Another common rhythm is that of germination, growth, pollination, fertilization, setting of fruit, dispersal of seed, and repeat. We rely on varying articulations of this general rhythmic pattern in growing food in gardens. The details change from plant to plant, and vary in different locations and according to season, but the general rhythm remains the same. Less well known but of perhaps greater significance are rhythmic phenomena such as the role of the lunar cycle in relation to plant growth. The connection between the movement of the moon and the ebb and flow of tides is widely acknowledged. Yet the influence of lunar gravity upon seeds, plants, and humans which are mostly composed of water is relegated to the discredited realm of folk knowledge and looked upon with skepticism by modern science. Biodynamic farmers and gardeners continue to plant by the phases of the moon in contemporary times following wisdom of many pre-modern cultures who used the lunar calendar to time agricultural work. The movement of the moon in relation to the earth exerts similar force over seeds and plants filled with water. In addition, the reflective luminescence of the moon provides extra light to plants at night. Together the ebb and flow of gravitational force and varying amounts of lunar light affect growing plants in a variety of ways, describing two interrelated sine waves. Biointensive and biodynamic gardeners recognize this. John Jeavons (1974), for instance, describes the specific effects of the lunar cycle on plant growth. According to him, the following general plant growth patterns emerge. During the first week of the lunar cycle following the new moon, plants grow in a balanced state as lunar gravity decreases and luminescence increases. As lunar gravity and luminescence increase during the second week of the lunar cycle, in approach of the full moon, leaf growth increases alone while root growth decreases. Following the full moon, in the third week of the lunar cycle, a decrease in both lunar gravity (and associated relative increase of earthly gravity) and luminescence results in enhanced plant root growth. The fourth and final week of the lunar cycle results in a balanced plant "resting period" during which leaf and root growth are curtailed. This is a good time to sow seeds. The rhythm of the lunar cycle repeats every 28 days in this predictable sequence.

Fundamental to ecological systems and all living things are weather patterns formed through the convergence of familiar rhythms expressed on grand scale. Warm moist air originating from the intertropical convergence zone sets broad patterns of global air circulation in motion. The regular movement of air in sine waves is influenced by varying degrees of atmospheric heat and forms upward heat spirals (observe soaring birds "riding" the skyward wave) which in turn carry moisture into areas of lower atmospheric pressure where it condenses and creates conditions for rain. The same phenomenon occurs when warm moist air encounters physical obstacles such as landmasses, continents, mountains, forests, and buildings. Ecosystems are significantly affected by these natural cycles. Weather patterns also affect landscape dynamics, such as hillside erosion or valley silting, as well as biological constitution, such as soil and related flora and fauna, and eventually even human culture. Indeed biological diversity embedded in varying geographical habitats supports and gives rise to cultural diversity, a dynamic interconnection discussed in further detail in Chapter 7.

Many cultural traditions are often associated with seasonal changes, and many indigenous cultures count time in reference to natural cycles as expressed in complex patterns rather than simplified in mechanical clocks or calendars. Cultural and religious activity is also associated with changing levels of light and seasonal variation. Zoroastrians, for instance, tie their sacred thread and have different prayers for different times of the day and face the sun while praying, bowing in humility. Rituals that are in tune with nature's cycles are practiced across nations and cultures.

Pedagogical Implications

Recognizing natural cycles as pre-eminent influences upon all aspects of life including human social activities, a general appreciation of different rhythms may be helpful in designing an education for enrichment of living systems. Tragically in most cases educational processes are set to the lifeless rhythms derived from a mechanistic orientation. Modern schools are generally modeled after industrial examples in which speed and size are always faster and bigger, respectively, and thought of as qualities of efficiency. In the name of efficiency and progress, school buildings are often consolidated to serve greater numbers of students; learning is tiered to linear developmental growth models; subject matter is arbitrarily divided by distinct class periods. Talk of taking successful school design "to scale" invariably means an increase in magnitude without regard to orders of magnitude; often something is lost. What works at one scale does not necessarily translate as scale grows beyond its optimal size. The dominant metaphors of educational reform at present refer to speed as an explicitly desirable quality: *Race to the Top* casts teaching and learning as a competitive event. To this we ask, to where are we racing in such a hurry? On the other hand, gardens communicate a life rhythm and pace of activity that is in tune with natural cycles and oriented to functional earthly scale; this is distinctly different than the dominant speed of modern Western culture, which presently guides education. We might say the pace of a garden follows "the beat of a different drum," apart from the dominant beat of modern culture, set to the lifeless tick of a mechanical clock, with a linear conception of time defined by economic and political "seasons" rather than ecological cycles. An experienced gardener knows that given the right care, seeds will germinate in their own time, fruits will set in their own time, and buds will unfurl when ready. The processes cannot be hastened.

Modern education remains in large part guided by an archaic factory model of teaching and learning that directly contrasts the life rhythms and scale of living systems. Ringing of bells signals stark beginnings and endings of learning periods framed within boundaries of a timed school day. Learning is constricted to these predetermined hours. When a student leaves a math class at the sound of the bell and enters a history class, mathematical thought ends and historical thought begins. When the school day ends, learning stops with the definitive sound of the bell and "real life" begins again. Teaching and learning are condensed into chronological linear time blocks which contrast the more natural rhythms of the human mind and body. By regimenting learning in this way, it serves the school schedule more than it does the needs of students. For example, a student who is really engaged with a science experiment is compelled to abandon the work station abruptly to take up the study of Shakespeare. Wondering as to what would transpire in the beaker, instead, they must now make a mental shift "to be or not to be" and engage with Hamlet. At the same time, a student who is engaged with Hamlet's questions is now forced to abandon Shakespeare and rush to a science laboratory to study the properties of sodium chloride, instead. With this arrangement it is no wonder that when they show up to class, students are not immediately interested in the subject orchestrated by the ringing of bells. How could anyone be? Worse yet, carrying on with mechanistic sequencing of time, students are being pressured to learn more in a faster amount of time, as if the organic process of life and learning could be accelerated like the motor of a car.

How can we apply observations of living systems to education? Turning away from using modern mechanistic metaphors as salient guide for education, ecological systems can provide a regenerative alternative. The artificial division of learning into linear time blocks is a problem: all subject areas are interrelated to one another and to "real life" in countless ways. The stark division of time into disconnected blocks serves to obscure this basic reality and reduce learning to mastery of the parts without recognition of the whole to which they belong. Unlike living soil, which smoothes and blurs boundaries between known and unknown, beginnings and endings, the mechanical rhythm of the clock forces schools into rigid starts and stops, which creates friction between living systems (i.e. human beings) trying to connect with one another (i.e. teach and learn). In addition, the scale of education continues to increase. Imitating business, large-scale mergers of school buildings to create greater "efficiencies" overlook the attendant anonymity of children and youth that tends to accompany vastness.

Examples from Learning Gardens

In order to bring life to schools, we suggest recalibrating our sense of time and space through discovering rhythm and scale. Working in learning gardens, in full-bodied engagement with living soil, we help students to tune into the natural rhythms that sustain life. A linear conception of time ever moving forward toward "progress" becomes inadequate in contact with the recursive and cyclical pace of living systems; similarly, the modern disposition toward "thinking big" (Berry, 1970b) encounters serious challenge when we make hand contact with living soil. Learning gardens show a pathway whereby education can be designed to accommodate human scale following the pace of life. In the garden we see a pace of growth and change that is slower than the rapid and mechanical race toward "progress" or "the future." In the coldest and darkest months of winter we plan for the coming garden season; we create fiber crafts like scarves and sweaters and we eat the dried fruit of summer, along with gathered nuts and preserved foods. We notice the new wildlife that frequents the garden and we see exposed bird nests in leafless trees. In early spring, we harvest overwintered greens if we are in more moderate climates; we pull back the mulch from our garden beds to warm the moist soil and sow early greens; we prune fruit trees; and in later spring we prepare the ground for planting summer crops. Classrooms come to life with trays of seed sprouts in windowsills. When the sun's heat proclaims summer, we transplant, begin to harvest, and anticipate preserving fruits and vegetables in jars, dry and pickle, and relax in the shade. In autumn we sow the garden beds to cover crops to protect soil from erosion and increase fertility; we collect fallen leaves and spread them over our garden beds; we plant trees. Meanwhile the salmon return from the oceans and spawn in the rivers and streams of their youth; in the school learning gardens children return from the pleasures of summer break and harvest pumpkins and build compost piles. As days grow cold we turn inward as the decreasing light of the sun makes room for the introspective light of consciousness; we gather our bounty together with friends and relatives and reflect on our place in the life giving order. There is a seamless congruence between learning and the spiraling cycles of life unfurling through seasonal time.

Noticing Rhythm and Scale in Gardens

Spiderwebs in a garden demonstrate a valuing of redundancy in pattern form. No two spiderwebs are alike, though they all follow a general weaving pattern. And while the individual strands of the web appear delicate, they are actually incredibly strong and resilient, able to hold captured prey heavier than the mass of the entire web. Critical to the construction of a spiderweb is relationship of scale and magnitudes of order. Similarly, the butterfly in a garden presents a relevant example that integrates rhythm, scale, patterns, and transformation. Students of all ages learn a lot from observing the different stages of growth of a butterfly; planting fragrant flowers in a butterfly guild can help attract these intriguing creatures to learning gardens of all sizes. There are four stages of growth that proceed in a predictable rhythm: egg, larva, pupa, and adult (butterfly). The transition from one stage to another cannot be hurried in the way that the stages of childhood are routinely quickened in modern times. The metamorphosis of the butterfly, and many other insects, amphibians, and animals, all follow *kairotic* or event-based time. The lessons of transformation that students can learn from the butterfly are hugely relevant to human development: that there is a rhythm to change in tune with larger natural cycles; that change cannot be hurried for convenience; that there is a time for different phases; that scale matters; and that beauty emerges from dormancy. Even in the short time span of a butterfly's life, students can learn a lifetime of lessons if they are taught to be alert to observe these phenomena.

Songs and stories, too, embed information within rhythmic timing. The use of song as one method of ancient oceanic navigation demonstrates the symbiotic relationship between content and form and emphasizes a recursive conception of human activity within repeating natural cycles. Children's handclapping songs are a good example of a living link between rhythm and content that persists in schoolyards today. We have seen teachers and students singing garden and nature songs that seem to revive in them the spirit of life-giving qualities of the garden. Further, in contrast to written history, oral histories, often related in song or to a given rhythm, similarly embed content within a form that conveys a recursive sense of time. It is natural for older people to carry significant experiential knowledge as they have been more times around the sun; oral histories honor the intergenerational link between the old and the young, describe a recursive spiral pattern that recycles cultural energy in a dynamic way, and recognize the timeliness and timelessness of certain forms of knowledge in relation to natural cycles or stages of *kairotic* human development.

Seasonal Calendar for Schools

As on a farm, on school grounds, too, seasonal garden calendars guide activities and learning. One month of a seasonal calendar of events from a Portland school is presented in Figure 6.2 (Taylor, 2009).

September 2010

SUN	MON	TUE	WED	THU	FRI	SAT
			1 — Set up school composting and recycling systems.	2	3	4
5	6	7 — Children return to school as the days begin to get shorter and summer's bounty reaches its peak.	8 — Introduce students to the school and class garden.	9 — Plan a garden scavenger hunt.	10 — Visit local markets for supermarket scavenger hunts.	11
12	13 — Create nature journals & find favorite flowers, fruits & vegetables in the garden.	14	15 — Harvest potatoes when the tops die off. Dig out a whole plant to study. *Support local farms; buy school lunch.*	16 — Students prepare snacks using food in the school or family gardens.	17	17
19	20 — Harvest winter squash when the ground spot turns from white to cream or gold.	21	22 — Collect flower seeds, make seed packets to give away or sell at Harvest festival. Save seeds for spring planting.	23	24 — Simmer marigolds, sunflowers, or coreopsis for one hour to make a simple plant dye.	25
26	27 — Visit local community gardens or farms to harvest fall fruits and vegetables.	28	29 — *Support local farms; buy school lunch.*	30 — Teach concepts of root, leaf, stem, flower, and soil as you harvest summer vegetables	31	

Now in Season:
Apples, blackberries, broccoli, celery, chilies, cucumber, garlic, green beans, peppers, rutabaga, squash, tomatoes, watermelon, zucchini

Join Portland's Fruit Tree Project: share or harvest fruit in Portland Neighborhoods (portlandfruit.org) 503-284-6106

FIGURE 6.2 Seasonal garden curriculum calendar

Learning gardens curriculum can be tied directly to the seasonal pace of the year and introduces students to learning sequenced in congruence with the fading and returning solar light. Making students aware of the changing angle of the sun as they progress through the school year is an important way to get them to be in tune with seasonal changes. As the calendar indicates, on returning to school in the fall, students take stock of their school grounds, assess what is still growing from the summer months, what needs to be harvested, and what is ready for composting as beds begin to be prepared for winter sleep or winter crops. Harvesting beans, potatoes, or squash is particularly intriguing because each plant requires different methods of harvesting. Students create nature journals and find their favorite fruits and vegetables in the gardens. These are ongoing activities that encourage astute observation. Students also learn to make simple dyes which itself is an activity that we have found children of all ages to enjoy. Even as they collect seeds, students are often engaged in critical thinking about hybrids and about seed preservation techniques. Each month presents an opportunity to connect with seasons and celebrate the cycles of nature.

Discovering Natural Rhythm and Scale

School gardens recalibrate a sense of time and space in powerful ways. Working with living systems, the edges are often blurred so as to obscure the difference between beginning and end. Plant growth, for example, is nearly imperceptible: one day, a few seeds are just germinating and poking their heads above soil; before long, the garden bed is filled with sprawling plants. After warm autumn rains, mushrooms appear overnight around the base of trees and in mulch piles. Compost confuses beginnings and endings, as once living matter is converted to life-giving soil. The point where life begins and ends is impossible to detect, and made more difficult given the short lives of many of the micro-organisms responsible for breaking down rotting biomass. Scale and space are also refigured in living soil. Looking at a tree, for example, we know that much of the biomass is below ground. Even after a tree is uprooted in a storm, most of the roots remain out of sight. How extensive and vast are subterranean branches of tree roots, or of earthworm burrows, or mole tunnels? We can imagine by looking up at the canopy of a mature tree. Looking upward at the canopy, we note that a single mature oak tree exposes several acres of leaf surface to the sun. The branching pattern of trees allows for optimal sun exposure using the least possible land area. This is an example of efficiency using natural rhythms and scales as a guide.

Children and youth can appreciate the patterns and the rhythm of living systems, engaging with the living soil of school learning gardens. Gardens are visited by a variety of birds and animals: hummingbirds, blue jays, sparrows, swallows, ladybugs, honeybees, aphids, and caterpillars all with different sizes and colors, all with different life cycles and rhythms. Digging in living soil, children notice that slugs are sluggish and earthworms are wriggly. In a garden, children learn about fragrance associated with different times of the day or the intensity of the sun. For example, jasmine has more of an intoxicating fragrance at night. Children are fascinated with watching the sunflower follow the sun during the day, or by noticing how flowers open and close in response to the movement and intensity of the sun. Sunrise and sunset are sacred reminders of the gifts of life. The natural rhythm of living systems is guided by the movement of the sun and earth and moon, and modulates over time. The length of day varies greatly across the time scale of a year, and biotic activity responds accordingly. A fluid responsive coupling between organism and environment allows continual modification over time, often in recursive fashion.

Seasonal Curriculum: Investigation and Practice

At Lane Middle School in Portland, Oregon, a seasonal curriculum is being used to weave together an ecological vision of education linked to the natural rhythms of the garden. This achieves two goals: the subtle changes of seasons from fall to winter to spring are highlighted and celebrated, and topics of study are enlivened by their apparent relevancy to present circumstances. A cyclical pattern of learning emerges as students spiral through grades and seasons. By correlating state benchmarks with seasonal activities, garden learning takes on more meaning as a central venue for integrating new concepts. An example of seasonal curriculum is outlined in Table 6.1.

Themes such as decomposition and energy transfer are taught in autumn, as this is the best time to build compost heaps. Because compost heaps are constructed each autumn, every year a new layer of learning is added. In younger grades, students have learned that decomposition is a scientific word for "rot." Now they learn about common compost ingredients, which are known as greens and browns due to their general colors. Greens represent the element nitrogen, and are found in grass clippings, food waste, and, the one color exception, manure. Browns are mostly made of the element carbon, and are found in straw and dry leaves. Students collect these ingredients and mix them in even amounts to create a basic compost heap. While they are building the heap, students notice worms, pill bugs, or millipedes scurrying around under piles of moist leaves. Many have noticed this at home when they help to rake autumn leaves. Lessons on the role of decomposers in the food web come alive when linked with actually building a compost heap. In the unit on "conducting investigation," students enjoy meeting fungi, bacteria, and insects—the "FBI"—that work tirelessly to transform rotting food into rich compost. They recall what they have learned in the science laboratory about nutrient accumulation and elements needed to stimulate decomposition. Scientific knowledge translates into actual practical knowledge when these adolescents learn how to stimulate decomposition with a mixture of greens and browns. Common garden wisdom specifies an ideal compost heap constitution of 30 parts carbon to 1 part nitrogen (30:1). Drawing upon their mathematics classes where they have studied ratios and adding ratios, students use lists displaying different carbon and nitrogen ratios of possible ingredients, such as leaves (50:1), straw (100:1), and manure (10:1), in order to determine what volumes of each ingredient are needed to find a near ideal ratio of 30:1.

Seed saving is another autumn activity that generates a number of meaningful avenues of academic inquiry and skills practice. For students, harvesting seeds involves many mathematics skills such as counting, estimation, and multiplication, as well as scientific concepts such as moisture density (Thorp, 2006). They learn that seed production is not just a means of reproduction; their diversity is also one way in which plants move themselves across the landscape year to year. Students observe the varying strategies that different plants have developed for moving themselves: some plants such as lettuce let the wind scatter dry seeds; others, such as pumpkin, surround their seeds with delicious flesh in hopes that an animal may eat the seed and later excrete it out in another location, conveniently fertilized. Students learn about reproduction, genetics, evolution, and the role of wildlife in plant reproduction. A Life Science unit on plant

TABLE 6.1 Seasonal garden curriculum

Grade	Fall Lessons	Winter Lessons	Spring Lessons
5th Adaptation and Diversity	• Seasonal change & adaptation • Observation & sensory evidence • Decomposition • Creating a model • Energy Transfer • Diagramming life cycles • Conducting investigation	• Fruits and roots (structure and function) • Dissection • Storytelling • Cooking and tasting • Photosynthesis • Dramatization of chemical change	• Flowers and pollinators • Dissection and identification • Labeling a diagram • Pollination • Creating a model • Interdependence in garden environment • Observation and classification
6th Life Science	• Observe and categorize plant density and growth	• Graphing • Two-dimensional geometry	• Flowers, structure and function • Insects
7th Physical Science	• Physical and chemical properties • Compare types of energy	• Physical and chemical properties • Name elements	• Two-dimensional geometry • Estimation
8th Citizenship	• Writing modes and strategies • Identify audience and purpose	• Advertising and persuasive writing	• Educational signage and descriptive writing

breeding is linked with the saving of seeds. Students at Chicago High School, for instance, are intrigued that plants grown from hybrid seeds display great vigor but cannot reproduce. This presents an area of inquiry that is itself a hybrid between science and social studies: why would people create hybrid seeds if the plant will not be able to reproduce? What are some benefits and drawbacks of plants grown from hybrid seeds? If farmers cannot save viable seed from hybrid plants, then who controls the allocation of seed? These questions frame a debate of contemporary global significance. Additionally, the historical and cultural information encoded in heritage seeds provides a living link between the sciences and social studies.

When the garden is dormant in winter, we find students at Lane Middle School in Portland, Oregon mapping vegetable beds, researching botanical, historical, and cultural facts about different plants, and creating a planting calendar. Winter is a good time to bring attention to the storytelling aspect of food and gardens. A number of schools welcome native storytellers; gathering in circles, students from San Francisco to New Jersey become contemplative about the cycles of life even in the winter months.

Learning the relationship between different plant parts is also appropriate during winter in preparation for the spring. In late winter or early spring, test sprouting beans in moist paper towels as seen in Chapter 5 is an easy and exciting activity that foregrounds outdoor planting and revisits the earlier study of plant parts. Following the last frosts of spring, seed sowing and transplanting commences, presenting opportunities to learn practical care of living plants.

Teaching About Scale: Square-Foot Gardening

When designing and planning the spring garden at the Learning Gardens Laboratory at Lane Middle School in Portland, Oregon, sixth and seventh graders are learning about square-foot gardening. This is used as a rubric to explore issues of scale by comparing how many seedlings can grow in a common area, one square foot. Even though plants do not grow in squares, this methodology is helpful in planning garden design. Since many of the students may have little or no experience growing a garden, they do not realize that a broccoli plant will take up more area than a pea plant or a carrot. Square-foot gardening consists of overlaying a 1 foot by 1 foot grid on garden beds and planning placement of plants according to their relative size. The picture in Figure 6.3 displays square-foot gardening.

Stakes and string are used when this is done in the field; alternately, on rainy days grid paper can be used as a basis for an indoor activity. Mature sizes of plants are represented in terms of the number possible per square foot. For example, a single broccoli plant requires an entire square foot area; in the same size block, we can alternatively plant four pea plants or 16 carrots. A garden teacher asks students to determine how many square feet of garden bed are needed to grow a

FIGURE 6.3 Laying out a square-foot grid to measure plant spacing

certain number of broccoli plants, or peas, or carrots. At more advanced levels, students are able to adjust the square-foot design to accommodate more nuanced shapes or integrated designs. At all levels this is a simple way to connect two-dimensional geometry scale with garden design. The students also learn about perimeter and area by actually measuring their plots.

Closing the Loop: Learning Gardens, Living Soil, and Sustainability Education

At present, modern education is tied directly to the chronological time and life-less mechanistic models guiding the modern economy. Schooling is described as a means to an end, alternately phrased as essential to develop job skills for the new information economy or as necessary to ensure global economic competition. Lost in this discourse is the value of life itself and the need to learn in such a way that honors life. Practical engagement with the living soil of learning gardens prepares students to encounter the social and ecological challenges of the twenty-first century in creative and sensitive ways. Carrying living soil into the classroom as a metaphoric construct asks us to discover natural rhythm and scale and tune our pedagogy to a life-enriching tempo. We need a more nuanced ecological vision of educational processes rooted in the timeliness and timelessness of natural rhythms.

A basic goal of sustainability education—and education in general—is enrichment of life and living systems. This can be achieved through following natural rhythms and creating social systems that are congruent with natural scale. Consider the moderating effects of garden rhythm on the pace and tempo of modern education. We propose learning gardens as a viable venue for conducting teaching and learning in a way that is less *chronic* and more *kairotic*, or, in the words of Jeffery Hawkins (2010), less timed or time-bound and more timely and time-generous.

Engaged with life work in the garden, one may lose the sense of time as a rigid boundary to be negotiated and gain a view of time as fluid and recursive. There is no set beginning or end point for garden activity: harvest represents as much the end of plant growth as the beginning of decomposition. A view of time as recursive or fluid mirrors an ecological understanding of living systems as dynamic process-oriented meta-organisms. Within ecosystems, no set beginning or end point exists; the rhythm of life does not lead to a certain conclusion, it is endless. Such a perspective may inform learning gardens pedagogy in particular and sustainability education in general. In terms of ecological sustainability, no set end point exists to which the activity of the present is meant to serve; rather, a confluence of infinite co-evolutionary feedback loops spiraling through time better describes the arc of a regenerative pathway.

7

VALUING BIOCULTURAL DIVERSITY

Diversity in nature and culture makes us human.

(Harmon, 2002)

I just love to see how even two radishes are not alike. Even the color is not exactly the same ... Everything in the garden is different. There is so much variety.

(Student, Learning Gardens)

The co-evolution of ecological and cultural systems is writ large upon landscapes and is particularly observable in the rich variety of soils. Biological and cultural diversity are intimately linked with one another, and it is difficult to neatly separate directions of influence; as such we are interested in students learning about dynamic and reciprocal interdependences among human activities and the environment. In this chapter we explore and discuss the relationship between biological diversity and cultural diversity.

Locating Biocultural Diversity in the Soils of Learning Gardens

Biocultural diversity refers to the inextricable link between biological diversity and cultural diversity. Learning gardens are uniquely situated at the symbolic and literal crossroads of biology and culture, and can be excellent sites for initiating students into lifeways more respectful of biocultural diversity. As food producing spaces, gardens are perfect locations to bring awareness to two critical points: non-human biotic community members have food needs, and human food needs extend beyond just the intestines. The first point is common knowledge; it is widely known that bees rely upon pollen for sustenance and that squirrels gather nuts in fall to support themselves through winter. The second point, however, is more nuanced and less well understood. In addition to food for gut, we also need food for thought and food for spirit. From this perspective, the cultural aspects of food—the traditions, the symbolisms, the taboos—take on as significant meaning as its nutritional value. Without reference to or awareness of this aspect of food, gardens are at best two-dimensional spaces dedicated to sustaining the visceral needs of human and non-human communities. Adding a cultural dimension to our understanding of food brings gardens to life as symbolic spaces connected with our inner lives, not just our inner organs. Perhaps we need a different word than "food" to express products of a garden oriented toward providing for the needs of non-human biotic community members as well as more-than-digestive food needs of human members. For example, gardens provide habitat and milieu for diverse wildlife, which are integral to growing food. As provision for non-human biotic community members as well as cultural sustenance, food symbolizes and directs awareness toward ways in which biological and cultural development are interlinked and woven together. Simply put, cultures do not materialize out of a vacuum but arise from the literal dust of food-producing soils through a complex co-evolutionary process interconnected with biological, ecological, and pedological systems.

Throughout human history, soils have defined human societies (Landa & Feller, 2010); for indigenous communities, soils are places of consciousness, not merely physical places (Cajete, 2001). Colors, textures, porosity, inorganic and organic elements of soils result from and interact with the diversity of human cultures. Since soil diversity and human diversity co-evolve in their interaction with particular places, culture plays a significant role in enhancing the diversity of pedons (units of soil). In order to conserve soil diversity and life, it is therefore imperative to conserve cultural diversity. With this in mind, we propose that reverence for soil and work in learning gardens act as pathways toward better understanding of biocultural diversity.

Ecological Perspective

An ecological perspective links biological diversity and cultural diversity in terms of interconnected systems and highlights their coupling as a dynamic multidirectional relationship. Culture is continually formed through human interaction with the living landscape; in turn, cultural mental models, beliefs, rituals, and practices influence and reshape the landscape. As biological and cultural diversity are intimately linked, the sustenance of one indeed relies upon the vibrancy of the other. The struggle for diversity is an effort that requires fostering links across a range of domains. The term "monocultures of the mind" (Shiva, 1993) poignantly highlights the critical connection between dominant homogenizing mental models or ways of thinking and diminishment of biological and cultural diversity.

Diversity advocates such as conservation practitioners and proponents for local/ indigenous communities can optimize effectiveness through locating productive ways of working together. The irony is that homogeneity threatens not only one group or another, but all of us; thus we can choose to put into place ways of remembering, ways of preserving, and ways of rescuing the earth's diversity in all its forms. Michelle Cocks (2006) explains the connection between biological diversity to human welfare and survival:

> It is important to explicitly recognize the role played by human diversity in biodiversity conservation because biodiversity represents a source of raw material on which the processes of evolution depend. The less diversity there is, the greater the chance that life could be destroyed through lack of resilience to environmental change. Biodiversity needs to be maintained because it provides humans with different ways of understanding and interacting with the world and ultimately offers different possibilities for human futures.
>
> *(p. 187)*

Similarly, Frederick J.W. van Oudenhoven, Dunja Mijatovic, and Pablo B. Eyzaguirre (2010) point out the ecological basis for the creation and maintenance of human culture, noting the significant role of interactions among ecology and culture in supporting biological diversity:

> Culture can be understood as an expression of the interaction over time between communities and their natural, historical and social environments. These environments not only satisfy people's material needs for food, fodder, water, medicines and other natural resources, but also provide the bases for ethical values, concepts of sacred spaces, aesthetic experiences, and personal or group identities derived from the local surroundings.
>
> *(p. 12)*

Among indigenous and tribal populations, folklore associated with cultivated and wild plants and animals and cultural practices such as ceremonies, dances, prayers, and songs are often an expression of their respect and awe of sacred sites. Contrary to popular assumption, "culture" is not the sole purview of non-Western and indigenous peoples. As Cocks (2006) indicates, "the concept of culture is multi-dimensional ... it can be related to specific lifestyles and dominant modes of interaction with the natural environment, and to specific aspects of behavior ... relationship between humans and their environment[s] is mediated by culture" (p. 191). We understand culture as a coherent system of values, beliefs, and ideas that social groups make use of in experiencing, making sense of, and modifying the world in mutually established ways. The links among biological diversity and cultural diversity are therefore present in East and West, South and North, rural and urban communities. Biocultural diversity and threats of homogeneity can be of interest to all people, no matter their place or location. In keeping with the significance of context as discussed in Chapter 4, we can begin valuing bio-cultural diversity where we are in place. Sustainability education, then, teaches about and is embedded in honoring diversity, particularly soil and linguistic and biocultural diversity. In showing the connections among linguistic, cultural, and biological diversity, Luisa Maffi (2005) explains:

> Biocultural diversity comprises the diversity of life in all of its manifesta-tions: biological, cultural, and linguistic, which are interrelated (and possi-bly co-evolved) within a complex socio-ecological adaptive system. ... The diversity of life is made up not only of the diversity of plants and animal species, habitats, and ecosystems found on the planet, but also of the diversity of human cultures and languages.
>
> *(p. 269)*

These diversities interact with and affect one another in complex ways; they do not exist in separate and parallel realms. Maffi (2005) points out that 50–60% of the 6,000-plus languages spoken are projected to be soon lost and replaced by larger majority languages (p. 602). She proposes that resources of many under-standings—both cultural and linguistic—be pooled in order to gain a variety of perspectives on how to address ecological problems. Furthermore, she points out, it is through mutual adaptation between the environment and humans at the local level—defined by place—that the links among these diversities have developed in a co-evolutionary manner. Biologically diverse places are also culturally diverse (Parajuli, 2001, p. 584), a truism which reflects the ways in which through differentiated cultivation and use of diverse plants, cultural and linguistic diversity contributes to proliferation of multiple species. As Jules Pretty and his co-authors (2009) explain:

> There is common recognition around the world that diversity of life involves both the living forms (biological diversity) and the worldviews and cosmologies of what life means (cultural diversity). The importance of this diversity is increasingly being recognized in industrialized societies and especially in urban areas where people are disconnected from their natural resource base.
>
> *(p. 101)*

The human species can no more dissociate itself from nature than it can divorce itself from the products of cultural creation, as "biological diversity and ecological processes are the anvil on which human physical and mental fitness are formed" (Kellert, 1996, p. 9). Convergence of biological and cultural diversity occurs at many levels and exists in dynamic feedback loops. One area where the connection is readily apparent is that of language:

> knowledge bases evolve with the ecosystems upon which they are based and languages comprise words describing ecosystem components. If plants or animals are lost then the words used to describe them are often lost from a language shortly after ... and this will change the way the natural environment is shaped by the practices and livelihoods of those human communities.
>
> *(Pretty et al., 2009, pp. 102–103)*

Diversity promotes resilience in biological and cultural systems. The problem of homogeneity is evidenced in the increased disease susceptibility of crops grown in a monoculture. Planting a diverse array of plants encourages biological feedback loops which act to balance agroecological systems. Homogenization of human culture and biological life are also linked. As diverse cultures are assimilated (often forcibly) into a dominant culture, both cultural and biological diversity are lost. In place of endemic plant species or varieties well adapted to specific cultural practices (such as annual timing of ceremonies), culinary sensibilities (such as palatable combination with other crops, or comparable time of harvest), climatic factors, soil conditions, or other specific qualities of place encoded in culturally diverse ways of knowing, a monoculture of the mind imposes a single species/variety crop regime and diet (Shiva, 1993). The danger of a monoculture is that it tends to be self-reinforcing in both mental and physical manifestation. As biodiversity contributes to natural resiliency, cultural diversity can positively impact the flexibility and adaptability of social systems.

Pedagogical Implications

Homogenization in an era of globalization seems rampant. The reports of ecosystem fragility that we hear are often linked with loss of languages, loss of cultural understandings related to people's relationship with the land and its bounties that evolved over centuries, and loss of biodiversity. As Cocks (2006) explains: "the very same processes of global socioeconomic development that destroy biodiversity also cause local cultures to be swallowed up in the expansion of the global economy" (p. 140). The loss of diversity to homogeneity is also seen in our educational system that promotes standardization rather than creativity. Similarly, homogenized school buildings on asphalted ground metaphorically capture the loss of variety in terms of not only curriculum but also pedagogy necessary to deal with the diverse needs of children and youth. An editorial by Kent Redford and J. Peter Brosius (2006) for the *Global Environmental Change* points out the consequences of homogenization:

Whichever forms of diversity we might wish to focus on, the drivers of homogenization are much the same: the hypermobility of capital, increasingly unfettered free trade policies driven by neoliberal agendas, structural violence and physical violence; all those things we gloss under the label "globalization." No matter the cause, this cascading homogenization produces the same result—a simplified world. In fact, in the biological literature a name for this impending epoch has emerged—the *homogocene*.

(p. 317 emphasis added)

In countering the trend toward homogocene, valuing biocultural diversity brings life to the center of the educational enterprise, and resists the simplification of the world to which children are introduced. Valuing diversity is reinforced by creating diverse garden learning environments.

In *The Earth Knows My Name: Food, Culture, and Sustainability in the Gardens of Ethnic Americans,* Patricia Klindiest (2006) captures the stories of ethnic immigrants who revive their memories and represent the cultures of their homelands through their gardens. In re-enacting the agricultural traditions of the people they have left behind, for the immigrant communities she talks to, "food is their favorite topic—the bridge that carries them over a river of loss" (p. 24). Visiting ethnic gardeners in Southern California or in Amherst, Massachusetts, Klindiest finds that for each gardener, the garden represents the strong connections they embrace and embody with their past and their culture, but most importantly, it symbolizes their deep allegiance to the earth. The intimate integration of food and culture in their stories also supports rejection of the dominant capitalist culture and economy.

Examples from Learning Gardens

One of the goals of learning gardens is to promote multicultural learning. Generally, in schools, multicultural learning solely addresses cultural diversity. However, in learning gardens, we see an opportunity to integrate cultural diversity with biological diversity.

Surrounded by concrete, brick, and asphalt, school gardens are islands of rich biological activity within a sea of modern edifices. This refers to more than sprawling squash plants and teetering corn stalks that announce entry into a plant-based world. Animals of all sizes and shapes also thrive in school gardens. Walking along mulched pathways, often we see many bees, butterflies, birds, and wildlife otherwise displaced from their natural habitats by development. More than food-producing sites, gardens are thus places of refuge for biotic community members evicted from their local residences of long tenure. Water features are particularly effective in bringing birds to gardens; a simple water bath can be enough to attract numerous species of birds to the land. Birds play an important role in controlling insect populations and brighten even gray winter days with their beautiful plumage.

Planting borders of flowers and other aromatic plants attracts bees, which are essential for pollination and fruit set. A simple sprinkling of white clover among conventional grass seed can be sufficient persuasion to bring gentle bumblebees in great numbers. Gathering and piling wood and stones at the edges of garden beds provides simple homes for snakes, which feast on rodents that may cause havoc near food production areas. These techniques and more invite the more-than-human community to share the garden with humans as a sanctuary among the frenzy of modernity.

In terms of biological diversity, the world that today's children are inheriting is greatly diminished from its previous vibrant splendor. This is not to romanticize an ideal time when nature was free of human harm; rather, it is to recognize the basic fact that while humans have always interacted with nature across time, many scientists are raising awareness of an alarming rate of plant and animal extinction (Kellert, 2005; Wilson, 1984). The rate of species extinction now exceeds that of discovery, which means that many species will be lost before they are even found.

School gardens can be effective preserves of local biological diversity, and can teach students about the critical importance of protecting endemic species for their own sake through nurturing natural habitat. Seen in this way, gardens are more than another site for the aggrandizement of human needs and desires as mere food production zones, but are at least in part areas created to restore balance to estranged relationships among human and biotic community members. Gardens thus provide sustenance to more than just human participants, can demonstrate that humans are but one strand of the web of life, and teach that our activities and installations can be beneficial to the community. A fruit tree, for instance, provides more than fruit for humans. It also gives pollen for bees, apples for squirrels, dappled shade for daffodils, crevices in bark for insect homes, root hairs for mycorrhizae bacteria, and leaf debris for worms. Non-human biotic neighbors have food needs that human gardeners can support with a spirit of gracious reciprocity.

Multiple Plant Functions: Activity for Students

Within biologically diverse learning gardens, many plants' principal function relates to sustaining non-human biotic community members. Plants such as comfrey, for example, which feeds bees through the summer, can be respected and valued for their contributions beyond humans. Students may learn that comfrey leaves can be used as a medicinal balm for human wounds and rashes, a lesson which raises awareness of the existence of non-food cultural uses of plants. Table 7.1 shows an activity rubric that students and teachers can use to consider the multiple functions and uses of plants.

TABLE 7.1 Food connections activity. For each crop or herb, collect content in these divisions as your reference materials from which to create curricula

[A] Ecological	[B] Agricultural/Gardening/ Culinary	[C] Cultural/Healing
Names • Common • Scientific • Other languages	Cultivars or Varieties • Native relatives • Human produced • Heirlooms	Literature • Folklore • Poetry • Stories
Botanical Information • Phylogeny • Structure • Classification	Planting and Cultivation • Techniques • Care requirements • Companion plants	Arts and Crafts • Plant as the medium • Plant as the subject • Plant as inspiration
Natural History • Origin • Ecology/Habitat	Growing and Cultivations • Watering and care • Soil preparation	Performance Arts • Music/Songs • Drama
Ethnobotany • Ecological names • Connections with insects/ birds/wildlife	Harvesting • Timing • Techniques • Saving seed	Festivals • Local • Regional • World
Science Opportunities • Chemistry • Biology • Drawing/Labeling	Preserving • Drying • Fermenting • Canning/Salting	Healing Qualities • Salves • Teas • Poultices
Range of Uses • Other than food • Traditional • Industrial	Cooking or Preparation • Recipes • Companion ingredients • Teas	Significance • Local Experts • Historical uses • Cultural relevance

References and Websites:_____

Trivia and Factoids:_____

The two-part insight that plants are more than food for humans alone contests the mechanistic industrial orientation of modern society and illuminates pathways toward more sustainable and compassionate human and ecological relationships. Promoting this relationship is critical, as Dennis Martinez (1996) warns: if you don't take care of the plants and talk to them and relate to them, they get lonely and go away. Several cultural non-food uses of plants include but are not limited to: tinctures, balms, dyes, fiber, building material, and musical instruments.

For example, Bindweed, *Convolvulus Arvensis*, vilified as a "weed," has roots and rhizomes that creep horizontally, producing further extensions that multiply with every tug of pulling off the plant. Its speedy growth can be a nightmare to farmers and gardeners. However, the plant has aesthetic value, and some gardeners build wire fences and trellises to prevent its spread and let the beautiful vine climb, to expose its morning-glory-type flowers. Jewelry boxes, decorating plates, and also textiles have depicted their beauty in art. The stem is quite strong and can be dried to use as a twine for plants in the garden. A pale yellow dye can be made from the leaves. Bindweed also has medicinal value serving as a mild laxative and diuretic. Some recommend using the leaves for washing spider bites. The roots and resin serve as purgatives. Another plant that students find fascinating is *Artemisia ludoviciana*, commonly known as White Sagebrush or Native Wormwood. Despite its name, the plant is not a sage (salvia). It has nitrogen-fixing qualities. Its yellow flowers can be dried on the stalk in late summer. The plant is used in Native American ceremonies and rituals to provide protection and spiritual cleansing. The indigenous ritual involves collecting the plant by offering of tobacco in prayer to one cluster of the plants to ask permission to gather medicine. The particular plant to which the offer is made is left untouched while gathering its neighboring plants in the garden. Tied in bundles and hung to dry, the sage is burned as smudge to purify the place before embarking on a spiritual discussion. The belief is that the sage helps with "washing" with smoke to aid people clear their minds and cleanse their spirits, readying themselves for the prayer. The highly fragrant plants are also made into pillows and saddle pads. Headaches, coughs, hemorrhoids, and stomach disorders are also treated with this plant. Innumerable opportunities exist for students to learn about the stories, myths, and varieties of properties of plants in the garden. When we understand that a given plant is more than "just a weed," we can begin to develop more life-enriching relationships between humans and nature.

Writing Poetry: Diversity in the School Garden

Allowing students to take the time to notice the variety of life-giving plants and flowers, to enjoy, and to express their admiration, wonder, and beauty in a variety of ways, keeps students of all ages engaged in schools. For instance, in Figure 7.1 we see the artwork that captures the variety of squashes that students draw.

In the following poem by a student from the Environmental Middle School in Portland, Oregon, one can almost hear the flowers and the bees and the intricate relationship of the adolescent with the natural and diverse world.

> *A conversation amongst the flowers*
>
> I had a dream that I could talk to the flowers
> That I could understand every fragment of every
> sentence of what the sunflowers said to me
> I could hear bees talking with the pansies, describing
> their day to each other
> I eavesdropped on a daffodil and a rose, who were
> whispering about the dandelion over next to the willow

FIGURE 7.1 First and second grade art: shapes, colors, and textures of squash

I exchanged jokes with a trillium while falling gently
into a pile of leaves that had fallen to the ground.
There, I found myself dozing off into a deep slumber
while the dogwood flowers sung a soft lullaby …

It is not uncommon for us to find students who are becoming aware of the variety of flowers and herbs growing in the garden. At his charter school in California, Krapfel (1999) encourages this kind of minute observation and attentiveness by taking his students to the same plot of land where they observe plants and are intrigued by their diversity.

Conserving Seeds, Seeding Memory

A logical first step that links biological diversity and cultural diversity in relation to plants concerns seed conservation and the maintenance of cultural memory through growing rare heritage and heirloom varieties. Through seed preservation and cultivation of heritage varieties, students can work toward supporting the Earth's fullest range of genetic biological abundance. Most learning gardens that we have been associated with teach students the value of seeds and seed collection, which is a compelling activity that conveys a strong sense of ownership of the garden. Easy-to-collect seeds that are child-friendly include radishes, sunflow-ers, corn, beans, pumpkins, and lupines. Seeds hold a crucial mix of biological/genetic and cultural information; conserving and passing this information forward fosters intergenerational reciprocity, celebrates biocultural heritage, and cooper-ates with local soil conditions. Laura DeLind (2006) reminds us that when we save seeds, we need to bank memories (cultural knowledge) not only germplasm; for her, "more essential and vital than banking, however, is the need to keep both (cultural knowledge and germplasm) in play" (p. 140). Similarly, Nabhan (1997) explains that science is neither a match nor a substitute for the nuanced know-ledge contained within native vocabularies, stories, and songs. Through creating a school-based seed bank, learning gardens conserve biocultural diversity in an effective and tangible way. Thus, one of the most important activities in a learning garden is to identify, feel, and collect seeds and then germinate them at the appropriate time. At one school, students germinate alfalfa seeds and then taste them and feel their delicate fragility. In another school, several students involved in seed collection learn about heirloom seeds harvested in their own school gardens. The various colors, textures, and shapes are not only intriguing, but are also inspiring as art objects. As DeLind (2006) indicates:

> Seeds contain a record of the needs, the decisions, and the environmental conditions that have come together and shaped any particular place and its inhabitants. They embody relationships and values just as surely as they carry genetic information. But to keep them (and the diversity they repre-sent) alive requires knowing the stories that accompany them and that explain or unlock their coded messages and meanings.
>
> (p. 140)

Through the growing of food, seeds and soil become active interlocutors between culture and nature (Anderson, 2009; Klindienst, 2006). How various cultures preserve seeds becomes an interdisciplinary unit that we have seen teachers explore across grade levels.

A Unit on Stone Soup

Another popular activity that engages students is "Stone Soup." Many adapta-tions of the folk story are now available throughout the world with myths that promote imagination: carrying nothing more than an empty pot, some travelers who visit a village are hungry. Since they do not possess anything, they ask for food. But the villagers, who themselves are not wealthy, are unwilling to share the little they have with strangers. So the visitors fill a pot with water, drop a stone in it, and place it over a fire. The villagers are curious about what the trave-lers are cooking. Upon investigating, they are told that stone soup is being made and they are invited to taste it. While the villagers find the stone soup to be good enough, they offer to add garnish to improve the flavor, which the soup is missing. One by one, villagers offer a diverse collection of vegetables—however little—resulting in a nourishing and delicious pot of soup. Not only does the story talk about the enjoyment of the soup itself with a variety of garden ingredients but the students learn lessons of cooperation and sharing. In varying traditions, in place of the stone, common inedible objects such as buttons, nails, and whatever is available in town, are added to the soup.

At Sunnyside Environmental School in Portland, students pick various vegetables from the school gardens and even bring some from home, in preparation of stone soup; as a shared experience, students of all ages chop up herbs to flavor their soup, which they cook in school and proudly eat with their families. While students learn about the diverse ingredients that are used to make soup, equally important, they learn the meaning of personal kindness, generosity, and trust. Besides learning and appreciating diversity, students also ask some basic questions about what makes them happy. How can they collectively solve the problems of hunger? Involvement in the story by identifying themselves with the characters motivates the students to seek knowledge and to discuss problems and search for solutions to the challenges of hunger and poverty in their own cities and communities. We have also observed and participated in "hunger banquets" that students host as an activity to honor Martin Luther King and his message of fighting injustice. The Oxfam America Hunger Banquet activity has become popular across the country in educating communities about global poverty and hunger. The basic arrangement is the random assignment of people who attend the "banquet" to meals that symbolize how different classes eat and access food. By actually experiencing the difference, an awareness is raised in the attendees about social and class injustices; discussions are held about how we might alleviate the problems of hunger in our communities.

Diversity Abounds in Apples: Writing Haiku

Third, fourth, and fifth graders at Sunnyside Environmental School in Portland, Oregon, involved in examining food from different cultures also research local and national varieties of apples and write haiku expressing the nuances of taste and texture that each variety embodies. While most people only know of three types of apples—red, green, and yellow—this activity introduces students to the astounding diversity of apples: Braeburn, Ambrosia, Blushing Gold, Cox's Orange Pippen, MacIntosh, Spartan, King David, Cortland, Buckeye Gala, Elstar, Mutsu, Granny Smith, Honey Crisp, and more. A sample of their haiku, capturing the variety of apples they studied is presented below.

The Sour Granny

Shiny green apple
Granny Smith is very tart

Makes my eyes water
Juicy and crunchy
Clean cut and the perfect snack
Sour, but refreshing

Cox's Orange Pippen
Laying on my plate
A watercolor painting
Green into red

Spartan
It's as red as blood
With stripes of green grass in it
After taste is pie

Cortland
Thick flames of sweetness
Yellow interrupts dark red
I'm full of sugar now.

The richness of the textures, colors, taste, and feel is captured in these haikus as students learn to notice subtle differences. Apples, thus, become more than a category of fruit.

Cultivating Multicultural Gardens

Learning gardens serve as an important connection to embodied learning and have enabled students to not only think about their own health but also that of the planet, resulting in a shift in the way they learn. Students in urban areas are particularly vulnerable to issues of hunger, obesity, and health given their loss of connection to how and where their food grows. In our experience of starting and supporting school gardens in Portland, Oregon, and also visiting several learning gardens in large urban school districts such as Chicago, Denver, Houston, San Francisco, San Diego, and Tampa, it is becoming increasingly clear that the non-white and/or poor and refugee students and their families who often live in either substandard housing or are on constant move are most likely to benefit from school gardens for a variety of reasons. A multicultural family garden at Fairview Elementary School in Denver, Colorado highlights ways that learning gardens are inviting venues for involvement of refugee and immigrant families. For almost ten years, Denver Urban Gardens (DUG), Slow Food Denver, and Learning Landscapes have partnered with Denver Public Schools to establish gardens at schools. Twenty of the 90 community gardens sponsored by DUG are on public school grounds. Working closely with families and children, DUG offers integrated nutrition and gardening support to school teachers and their classes. At Fairview Elementary School, a fifth grade teacher serves as the key staff person who has embraced the learning garden created on the school site. He has been actively involved and is invested in the school garden and integrates garden activities in the curriculum. With about 250 students, 96% of whom are non-white, the students at Fairview learn the significance of the metaphor of rainbow as they are encouraged to connect with the vegetables they planted over the summer with their families. They are actively involved in tending the gardens and to "eat a rainbow." As their teacher explains, he is able to connect knowledge and hands-on learning from the school garden to the science and math curriculum. Since there has been an influx of Somalian refugees in the school's neighborhood, the multicultural family gardens also invite the families to plant and harvest and encourage them to teach others about their homeland culinary traditions. Similar garden programs that value diverse families are to be found in public school districts in Portland, Oregon, and Houston, Texas. The diverse soils of school gardens also teach about honoring human and cultural diversity.

Closing the Loop: Learning Gardens, Living Soil, and Sustainability Education

Soil teaches us that diversity is the essence of life; in a single handful of living soil we discover a variety of organisms including bacteria, fungi, protozoa, nematodes, arthropods, and all sorts of decaying biomass. While many of the organisms are a mystery and are as yet unknown to scientists, what we do know is that soil is living (Williams & Brown, 2011). Living soil, being key to land's memory and cultural memory, must also be counted in the equation of diversity as value since culture, agriculture, and cultivating the land are all connected in significant ways. Diversity is not only a value of life, it is life. Beyond the norms of standardization in schools, sustainability requires the flourishing of differences. In weaving together biological diversity and cultural diversity through growing food in gardens, the empirical links between landscape and mindscape become apparent. As mental models affect human design of the physical landscape, so do the unique and varying nuances of living landscapes that shape and reshape the human imagination in a direction conducive to sustaining life. Gardens on school grounds dynamically animate the landscape with diverse life. Learning gardens are embedded in honoring diversity—particularly soil and linguistic and biocultural diversity. The living soil that sustains learning gardens is itself a rich source of diversity and grounds our understanding of biocultural diversity in the earth. In reminding us of the moral imperative of diversity, David Harmon (2002) captures it best when he pleads:

> The human species did not evolve in a world of drab monotony. Our brains, the consciousness function they produce, and the cultural variety

expressing that function have evolved over millions of years within a lavish, developing environment of biological riches ... our evolutionary history teaches us that cultural diversity is intimately related to the biological diversity of the nonhuman world. Current events tell us they face the same threats. The only effective way to meet them is with a cohesive, biocultural response. Through it, we would find that unity does not require uniformity.

(Harmon, in Maffi, 2005, p. 66)

Biological and cultural diversity cannot be separated one from the other. In learning gardens, there is ample opportunity to integrate biologically diverse species with culturally diverse understandings of plants. Nurturing interdependencies among biological and cultural diversity is foundational to creating and fostering favorable conditions for any form of sustainability.

8

EMBRACING PRACTICAL EXPERIENCE

Anybody can dig a hole and plant a tree. But make sure it survives. You have to nurture it, you have to water it, you have to keep at it until it becomes rooted so it can take care of itself.

(Maathai, 2006)

I get to work the soil and plant. It's hands-on instead of talking about it, I get to dig and get messy. That's my favorite thing.

(Student, Learning Gardens)

It is widely recognized that not all students flourish in a didactic, abstract, and reading- and writing-centric learning environment; many children and youth integrate new information best through practice or bodily engagement. The planning and planting of a school garden presents multiple opportunities to engage diverse ways of learning and brings all students into the conversation about life. Experience with the real world teaches us in profound ways. Active engagement, embodied learning, and practical experience are foundational to transformational education. In this chapter, we embrace practical experience in learning gardens as a dynamic form of engagement that animates teaching and learning and brings life to schools.

Life's Teachings in Soil

When children come into close bodily contact with soil in learning gardens, when they can actually feel soil "tilth," when they can personally compare the relative presence or absence of earthworms in different soils, issues of sustainable techniques are grounded in physical reality. The following sixth grade students' comments describe the educational value of learning gardens as such:

> It's about soil and science. We get to experience things. In class you get paper...

> You learn about animals and plants, how to harvest, how plants take time to grow. [It is important] not just to go to the store and buy stuff. You get to see and know how plants grow...

> In the gardens, we learn how not to fertilize, because it is not healthy to eat stuff that is artificially fertilized, it is healthier to eat organic.

The tempering and transformational qualities of experience are critical elements of an ecological perspective. The following eighth grade student's poem relates insights into the interconnected web of life that animates learning gardens.

Life

> The sun powers the plants
> The plants are used by animals.
> The animals are used by us, humans.
> But when we die our bodies belong to
> the earth.
>
> If you are not careful and you destroy
> one of these things you destroy the
> things that are in this cycle.

Insights such as those captured in the comments and poem above do not emerge from mental understandings alone; they are rooted in and nurtured by living experience.

Experience is not just about doing or hands-on learning. The ability to discriminate, make meaning, and learn is contingent upon pairing experience and reflection; this can teach a number of different ways of relating with life. For example, the same frost that kills some plants, such as pineapple sage, will ripen the fruit of others, such as persimmon. The occurrence of frost is neither good nor bad in its own right. Experience with it, paired with reflection, generates meaning and stimulates further questions such as: Why did one plant die and not the other? What makes persimmon ripen with frost, whereas other fruits need warmth to ripen? This experience presents a problem and contradicts what until this point might have been a taken-for-granted assumption that all fruit ripens with warmth; or that frost kills all plants.

The present experience contrasts previous experiences, stimulating questions about life, and, as Dewey (1938) puts it, "the principle of continuity of experience means that every experience both takes up something from those which have gone before and modifies in some way the quality of those which come after … the process goes on as long as life and learning continue" (p. 35). This is the phenomenon of "learning from experience." Present action is informed by previous experiences, while it at once reformulates knowledge and understanding that is carried forward in a transformed state. Recognizing the ripening effect of frost on persimmons, we may not hurry to pick them prior to a coming cold spell. Doing so, we carry forward our new knowledge, yet are still ready to receive new information that again unsettles understanding: the following season, if we leave the fruit on the trees late into autumn, we may notice squirrels harvesting the ripened fruit before we are able to. Interestingly, we also realize that if a hundred persimmons are likely to ripen within a month, there is no way that an individual can possibly eat them. So, we mimic nature and pick some unripe persimmons, bring them indoors, and when we are ready to eat, create an artificial frost by freezing and then thawing them indoors. The example of engaging with persimmon describes how a dynamic convergence of experience, discrimination, and reflection creates openness to new learning and change of perspective.

There is a degree of discomfort associated with challenging or rethinking previously taken-for-granted assumptions; all learning starts with encountering a "problem" (Dewey, 1910). The manner in which we approach the problem is what matters in terms of learning and growth. While encountering a problem may lead to rigid stubbornness and resistance to change, as Hans-Georg Gadamer (2001) points out, this need not be so:

> being experienced does not mean that one now knows something once and for all and becomes rigid in this knowledge; rather, one becomes more *open to new experiences*. A person who is experienced is undogmatic. Experience has the effect of freeing one to be open to new experience … In our experience we bring nothing to a close; we are constantly learning new things from our experience.
>
> *(pp. 52–53, emphasis added)*

In this light, the ability to discriminate is a motivator for learning, change, growth, and transformation; this view naturally contrasts the popular notion of discrimination associated with narrow mindedness and prejudice. Experience in the present moment takes us beyond pre-judgment. We need to let experience free us to the possibility of change; to acknowledge that there is no end to learning. Practical physical engagement with living things provides a wealth of opportunities to grow and learn.

Ecological Perspective

One intersection between mind and body is found in practical experience, where mental models meet physical reality. No one lives in mindscapes; all people work, play, grow, live, and die in physical landscapes that can be deeply known through experience. Poet Annie Dillard (1999) expresses the sentiment as follows:

> We live in all we seek. The hidden shows up in too-plain sight. It lives captive on the face of the obvious—the people, events, and things of the day—to which we as sophisticated children have long since become oblivious.
>
> *(p. 140)*

Experience brings us back into engagement with the social and ecological environments of which we are a part, into what Dillard (1999) describes above as the "face of the obvious." Practical engagement reconnects mindscape and landscape in a two-way feedback loop. Personal and cultural ideas, beliefs, and values influence understanding of the physical landscape and guide interaction; likewise, engagement with physical social and ecological reality shapes the establishment, growth, and change of mental models. Lacking practical experience with reality, the engagement feedback loop is blocked. For instance, without some degree of physical connection with the inner workings of soil, issues of chemical pollution or erosion are remote and abstract, as are protection and conservation techniques.

In the modern context, schools represent a promise of a "better life," which often signifies an escape from things manual. As an instrument of modern economy, schools privilege and celebrate the mind, while hand and manual labor are considered low-status and to be avoided; education is closely linked with economic and social development and notions of human progress. The 3 Rs of reading, 'riting, and 'rithmetic codify a privileging of the mind and marginalize manual ways of knowing. Moreover, the 3 Rs overlook the value of the 3 Hs of head, hand, and heart, eloquently articulated by Gandhi (1953). The 3 Rs are embedded in the first H, thereby expanding the discourse on learning to include the hand and the heart. Learning by doing, thus, is elevated as a legitimate form of inquiry and education as the head, hand, and heart are equally valued, and are conceived of in relationship to one another. Like individual members of a healthy ecosystem that interact and form dynamic and fluid interdependencies, head, hands, and heart sustain one another. Using all three faculties during experience, it is as if the three parts of the body are engaged in an ongoing dialogue. For example, when digging garden beds, the feel of the shovel stimulates hands, which in turn communicate information to head regarding soil condition, moisture, presence of rocks, and level of fatigue. The head contributes new strategies to optimize the task at hand; the heart may at this moment join the conversation, adding an ethic of care to the work. There is open communication among the parts; none are excluded from making sense of experience.

Importance of Experience: Dewey and Gandhi

Almost a century ago, on two different continents, Dewey and Gandhi proposed theories of education that overlap yet are based on different premises about modernity. While Dewey believed in the power of progress as embodied in a democratic social order, Gandhi cautioned about the unexamined consequences of "progress" and instead proposed an educational system that was grounded within the pre-colonial historical and cultural contexts of India to promote the simple, the non-exploitative, and the self-reliant life. While their goals of education related to the political and economic outcomes differed, for both men, learning by doing, relevance and context, and engagement in and with community were critical. In Gandhi's educational philosophy of 3Hs—head, hand, and heart—we find the connection with Dewey's philosophy of experience and learning by doing. Hence, here we examine what is meant by "experience" in education and how the two philosophers of education championed its role in learning. We begin with exploration of Dewey's concept of experience since he has elaborated the connection between experience and learning in several volumes: *Democracy and Education* (1916), *How We Think* (1910), *Experience and Education* (1938), and *Experience and Nature* (1925).

Dewey uses the principles of continuity and discrimination to determine the quality of the experience, since not all experiences are necessarily educative. In fact, some can be mis-educative. For Dewey (1916), to "learn from experience" is:

> to make a backward and forward connection between what we do to things and what we enjoy or suffer from things in consequence. Under such conditions, doing becomes a trying; an experiment with the world to find out what it is like; the undergoing becomes instruction—discovery of the connection of things.
>
> (p. 140)

Thus, experience and education cannot simply be equated. In the experiential continuum of education, only experiences that lead to growth and learning are to be pursued.

> The nature of experience can be understood only by noting that it includes an active and a passive element peculiarly combined. On the active hand, experience is *trying*—a meaning which is made explicit in the connected term experiment. On the passive, it is *undergoing*. When we experience

something we act upon it, we do something with it; then we suffer or undergo the consequences.

(Dewey, 1916, p. 139, emphasis in original)

Not all activities necessarily count as experience, says Dewey. It is the reflective connections between an action and consequences and then further action that builds on the previous experience that count as valuable. As he explains, experience as "trying" can involve change; yet change in and of itself is meaningless unless it is "consciously connected with the return wave of consequences which flow from it" (Dewey, 1916, p. 140). It is the significance of reflection that brings meaning to experience. For instance, in Dewey's example of a child sticking a finger in a flame, learning arises when the child makes the connection between the dangers of the action and hence in future not doing so in order to avoid a burn. In a learning garden, similarly, children may make connections between different treatment of plants and associated results. What happens, for example, when we protect some plants with glass cover and not others? Which plants require such protection, and which ones can survive the cold without it? The activity of covering plants before a frost does not become meaningful experience unless it connects with reflecting on the consequences and linking it with future action. As a sixth grader writes in her garden journal:

> We learn what season it is and what to plant in what season. I learned not to plant certain things in winter because it'll die.

Experience, thus, is not simply about active participation in learning as is traditionally construed. It is in the dialectic of action, reflection, and future guidance linking them in a continuum that brings about learning. Sound educational experience involves, above all, continuity and interaction between the learner and what is learned.

Dewey (1938) also cautions that the belief that "all genuine education comes about through experience" does not mean that all experiences are genuinely or equally educative. For example, consider a student who becomes interested in compost. Usually the experience of piling rotting materials and stirring with pitchfork is initially unappealing to students, but often after a few minutes of active engagement they are proud of their abilities and eager to learn more about what is happening within the pile. But digging and piling alone do not constitute an educative experience. Compost is best introduced experientially, and students' understanding of the process can be enhanced by further reflection following the activity. Not only does the quality of the experience matter, but so does what the student learns from it: future compost activities could build on previous experiences and new knowledge. "Every experience lives on in further experiences. Hence the central problem of an education based upon experience is to select the kind of present experiences that live fruitfully and creatively in subsequent experiences," leading to growth, according to Dewey (1938, p. 40).

Discriminating among and connecting disparate experiences requires reflection, that is, being able to discern relationships between activities that students are involved in. Unlike the overly didactic learning that usually occurs in traditional classrooms, direct engagement and participation with the real world is effective when students wonder about and connect the present with the past (what they know) and what the present might mean. A third grade classroom, for example, was involved in converting a vacant trashed school lot into a garden. The depth of reflection that students capture about their learning is seen in their journals. For example, this excerpt from a third grade student's journal captures the joys of working with living soil:

> We started collecting dirt inside a wheelbarrow then we dumped it in flower beds. We started digging little rows but we could not dig the rows too deep or else the plants might drown. It was fun working with the wet soil. It was not mud but it was still wet; it stuck to our hands.

Hence, reflection is about meaning-making. Because an experience is an interaction between a student and the environment (gardens in the above examples), there is change in the self and also in the environment. The change encompasses both the learner and the social and environmental milieu each impacting the other in profound ways. To be educative, students perceive meaning using the thought process. This point is significant since an oft-understood meaning of experience is simply to go outdoors and be involved in "hands-on" activities. Outdoor education and environmental education can fall prey to the same ineffective methods if they are merely activities disconnected from texts and previous experiences. Hence, school gardens are most effective in as much as how they are integrated in the curriculum and students can discuss, reflect on, and share what their gardening experiences are about. It is in the process of reflection that Dewey's "reconstruction and reorganization" of experience bring meaning to it. For instance, in Chapter 2 (pp. 18–19), a student vividly describes the transformation that she experiences when she and her classmates convert a trashed lot on school grounds into a learning garden. Writing about creating the garden, she reconstructs and reorganizes the experience and brings meaning to it. Reflection is a conscious, deliberate activity, not merely haphazard mulling over action. It can be taught and it can be learned. Embracing experience, thus, is more than about "doing" or a "technique" of teaching. There is intentionality in how teachers can use each active engagement with the garden to assist students in reflecting and meaning-making. Learning becomes a continuous process. The continuity of doing, learning, reflection, and experience enriches the educational enterprise.

Gandhi, like Dewey, emphasizes the "hand" as a symbol of active engagement with real life. While he is known as a spiritual leader who advanced the

philosophy of non-violence and *satyagraha*, Gandhi is little known for a fundamental theory of education detailed in *Nai Talim* or New Education. Rejecting
modern educational systems tied to the industrial economy and its vagaries,
endless wants, exploitation, consumption, and cultural domination, Gandhi
instead draws upon the vernacular wisdom of India to revitalize her villages,
her crafts, her economy, and her agriculture. In requiring the integration of
learning and living, Gandhi's educational program, like Dewey's, requires active
engagement with real life. While acknowledging the importance of the 3Rs, his
education of 3Hs seeks to give new value to manual labor (hand). It stems from
his concern for social justice and elevates the use of hands to high status, by
reconnecting individuals to their soils. By emphasizing the dignity of labor,
Gandhian education teaches self-respect and self-confidence to people whose life
and culture emerge from soil and are intricately linked to soil. Like Dewey, he
struggles against the contemporary notion that work is something we do after
graduating. Since work is an essential part of living, it must be integrated with
learning (Williams, 1993, pp. 18–21). Real life itself serves as the medium for
experience and learning. Horticulture, dairy farming, spinning, and weaving are
all undertaken with elders in intergenerational learning, embedding and rooting
schools in their local communities. The experience of living and learning is also
played out in service in Gandhi's educational program, just as the antecedents of
the present service learning movement in the West can be traced to Dewey's
experiential learning in community.

Pedagogical Implications

Modern schools tend to use didactic approaches to teaching and learning that
minimize the role of experience in learning and conditions children and youth to
be passive recipients of information. This pedagogical orientation reflects the
social and cultural milieu where: direct life experiences are mediated by technology and engagement is systematically replaced by vicarious experience (Bowers,
2000); unstructured and free play in natural outdoor settings is prohibited by a
climate of fear of the unknown (Louv, 2008); manual labor is marginalized and
devalued; environments are degraded; and the variety and possibility of life is
reduced to homogenized predictability.

Learning by doing, in place, over time, is important for nurturing an ecological balance between and among human cultures and biotic communities. Yet it
is not necessary to learn everything anew from personal experience alone; the
combination of firsthand familiarity with collective cultural or community
knowledge is foundational to an ecological perspective. The erosion of natural
experience as a hallmark of childhood coincides with increasing commoditization
of social and ecological commons. Stories, songs, knowledge, practices, and
customs contribute to meaningful engagement and communicate experience
intergenerationally. Authenticity in place emerges through practical experience

and connection with others in community. The link between mindscape and landscape is facilitated via functional relationships among human ideas, actions, and feedback from the land.

Examples from Learning Gardens

Learning gardens directly on school grounds are accessible to all children in all locations; experience with natural systems can be gained right outside of the school building. While soil and natural areas were once commonplace to childhood, a combination of changing landscapes and cultural values now distances children from the natural world (Louv, 2008). The land surrounding schools in many urban areas represents some of the last undeveloped soil. When these lands are liberated from artificial turf, chemically treated lawns, and impervious asphalt, learning gardens on school grounds come alive with biotic and human activity. Opportunities to develop practical experience and skills abound, along with possibilities for close engagement with living systems. There is a quality to learning through such intimate engagement with real life that cannot be equaled in book learning alone. By the same token, without a point of reference in curriculum, experiential learning can also be insufficient and ineffective. In learning gardens established directly on school grounds, experience stimulates a two-way search for answers to endless questions, and generates a back and forth movement from outside to inside learning. The answers to book questions lie in the soil, while many of the soil questions hide their answers in books. Moving between the practical and the abstract, learning is enlivened and takes on deeper meaning than through either method alone.

Camouflage and Wildlife

Children can gain experience engaging with wildlife in urban learning gardens which they may previously have overlooked. Seeing and hearing life in this way can motivate a search for answers from a textbook in a dynamic way. At the Learning Gardens Laboratory in Portland, Oregon, sixth grade students from Lane Middle School asked a teacher to help them use a classroom field guide to identify a bird they had noticed nesting in a pathway. Describing the bird's appearance and nesting habit, and with the teacher's guidance in reading the field guide, the class determined the bird to be a killdeer. Students then learned more details including its camouflaging features. The students, who had recently completed a unit on the use of camouflage in nature, began to understand the true meaning and application of camouflage as they discussed how difficult it had been to even notice the killdeer among stones.

In this case, practical experience stimulated questions that led to engagement both outside and inside the classroom. When learning moves back and forth from practical to abstract knowledge, a recursive combination of the two ways of knowing enlivens learning and creates deeper meaning. After reading about the vulnerable nesting habit of the killdeer, children acted to protect the area from foot traffic using bamboo stakes and string. They eagerly waited for the four eggs to hatch, and wondered where the male bird was hiding. Practical engagement with the nesting bird concretized the experience of the emerging spring; if the bird is laying eggs even in the cold, the students reasoned, perhaps it knows something of the coming warm weather. Building on this observation, they noted that the appearance of the bird could be correlated with planting schedules. Moreover, students concluded from reading about, and watching, that the killdeer favors rocky outcrops in selecting a nesting site. They wondered if creating a designed rocky place in a protected area might help more birds of this type nest at the learning garden. Thus, for the students, practical firsthand experience acted as a stimulus for learning that carries back and forth from outside to inside the classroom.

Culinary Connections

The process of planning, planting, and growing vegetables for use in school cafeterias or for donation to local food banks is another activity that relies upon and fosters experiential learning. For instance, when designing the garden, students consider the growth habits of different types of plants, along with the culinary possibilities offered by different species. For some, this may be the first time digging and planting in a garden, while others may draw upon previous family experience in choosing plants or designing garden beds. Either way, the experiential component is often lacking from education, thus precluding some from gaining practical skills and marginalizing the know-how of others. When planting a garden, some students gain new experiences, while others call upon past experiences that would otherwise remain dormant. The following excerpt from a teacher's journal (Anderson, 2009) describes how students draw upon past experience and integrate it with what they are learning in school gardens:

> Lucia talks about planting with her dad. "We planted tomatoes, cherry tomatoes." She demonstrates how to plant the tomato start. Watching her, how she delicately handles the fragile little plants lovingly, carefully, it is clear she has done this before. She shows us the knowledge she has learned from her dad. She gently handles the roots, separating the fine threads and placing them into the hole. "Dad says to keep the dirt with the plant. It's full of nutrients."
>
> *(p. 38)*

Students are most interested in exploring and harvesting the produce (Figure 8.1) and also picking and displaying flowers (Figure 8.2) that they have grown.

FIGURE 8.1 Gathering salad ingredients

FIGURE 8.2 Teasing with flowers

Students also learn about weather as it takes on new meaning when they are in touch with the sun and its energy. Mathematics comes alive as children count and weigh harvested produce, or plan the garden design. Geometry is grounded in the physical shapes of living plants and the relation between objects in the garden. For example, when students build a simple tripod trellis, they transition back and forth between abstract knowledge that triangles are stable geometric shapes and common sense, constantly mediated through actual physical experience. In some learning gardens, surplus food is donated to local charities, and older students are invited to participate in facilitating soup kitchens. While serving in a soup kitchen, one 13-year-old student from the Environmental Middle School in Portland, Oregon, states:

> Whenever I go the soup kitchen, there is always a long line of tired-looking people, sometimes with all their belongings slung over their backs ... white stubble growing out of the men's chins, shivering in the cold, waiting for food. They are clearly in need of a good meal, and if they could afford it, they wouldn't be there ... When we serve, we show people that we care about them. When people know that someone cares about them, they are generally happier and it gives them hope.

To another student who is also 13:

> food that we grow in learning gardens shows us how to be patient and also shows us the miracles of life. But, it also gives us skills so that we can learn to become independent and not always rely on large businesses to feed us. I think about the hungry and the homeless and how I can be of service to them.

Closing the Loop: Learning Gardens, Living Soil, and Sustainability Education

As the modern world becomes more "developed," the vagaries of vibrant living experience are homogenized and replaced with sterile vicarious or passive experiences. Neither in society nor in school is this adequate for engaging with and nurturing life and living systems. Engagement is a key element of sustainability education. Sustainability by any form or by any name cannot be experienced vicariously; moreover, creating a regenerative basis for human–nature relationships relies upon engagement. Practical experience with living soil in learning gardens provides a "growing medium" for learning principles of meaningful and reciprocal engagement with life. We understand life best when we engage with it; conversely, we can understand how to engage through interactions with real life. Practical experience is therefore one way to bring life to schools. Moving back and forth between previous ways of understanding the world and encountering new information through experience stimulates learning. This requires an openness to experience and an ability to discern and discriminate among experiences through reflection. Learning gardens can facilitate both access to and reflection upon experience directly outside the school walls. Growing food in living soil on school grounds exposes children and youth to innumerable real-life practical experiences and skills and brings up provocative questions for further inquiry. Learning through practical engagement with food, soil, and gardens goes beyond the particularities of specific plants or techniques. Learning comes alive through experience; at the same time, new experiences stimulate curiosity and further learning.

9
NURTURING INTERCONNECTEDNESS

> Ultimately ... there are no parts at all. What we call a part is merely a pattern in an inseparable web of relationships.
>
> (Capra, 1996)

> We are growing things we need together and I think it creates a great bond between people. A lot of learning [happens] in the garden. We are really distant from people that we live around, specifically in problem areas, we are extremely distant. [The garden] brings a community together.
>
> (Student, Learning Gardens)

We use living soil to highlight the life-giving order of interconnected systems as a model for bringing life to schools and schools to life. Vibrant living soil miraculously transforms death and decay into life and growth and continually recycles water and nutrients; these processes are facilitated by the interactions within complex and diverse micro-ecosystems which reside in the soil community (Hemenway, 2000; Hyams, 1976; Jeavons, 1974). All human cultures are inherently dependent upon such vibrant living soil, the foundation of terrestrial living systems. In this chapter we elaborate insights on interconnectedness from the study of living soil into pedagogical applications via learning gardens and sustainability education.

Interconnectedness as Relationships

Interconnectedness describes the existence of multiple and varied relationships that knit living ecological systems together. Other words that can be used to describe this essential characteristic of phenomena include "interrelatedness" or "interdependence." Beyond specific words, the salient point involves relationships that are multidirectional, dynamic, and ever-changing. When we observe nature, we see that all living things are connected even when the exact links are invisible or hidden from view; food webs and guilds are well-known examples of interdependence. Similar observation can be made of social systems; we perceive patterns of relationships that involve multidirectional symbolic communication such as teaching and learning, dynamic material transactions, and ever-changing networks of association such as families, schools, businesses and organizations. What in nature is referred to as an ecosystem is in social systems known as a community; interconnectedness is in both cases a central feature of system organization.

An appreciation of interconnectedness as a defining characteristic of nature, along with a realization of our inherent interdependence upon one another, and with nature, can have significant positive impacts with regard to our behavior toward others. Understanding our role as part of the web of life binds our human health with that of endangered and polluted ecosystems. Recognizing the social dimension of interconnectedness may inform more peaceful and compassionate ways of relating to one another across difference and in conflict. Neither of these hopeful outcomes is inevitable; certainly many forms of interconnection are negative, such as parasitic or abusive relationships. Our specific interest in this chapter is to advance education which recognizes and teaches about interconnectedness as a basic characteristic of all living social and ecological systems. Through positioning living soil as a central example of interconnectedness we highlight the life-giving possibilities of healthy ecological relationships.

We can choose to move from one-way relationships of dependence and exploitation into a symbiotic multidirectional relationship of interdependence with the soil community through returning organic wastes to the earth, refraining from using toxic chemicals, or self-regulating use of renewable resources. A good model for establishing such symbiotic relationship is that of nitrogen-fixing bacteria living in the roots of leguminous plants. Nitrogen is fundamental to all life. While it is in the atmosphere, plants cannot breathe it. And, it is scarce on its own in the soil. There are only a few ways it can be made available to plants, one of which is via a symbiotic relationship between the nitrogen-fixing rhizobial bacteria living in the roots of legumes and certain other plants. This mutual partnership ensures soil fertility. Understanding this relationship in literal terms empowers us to contribute wisely to build soil; if we carry this knowledge into the metaphoric realm, we realize quite profoundly the necessity to acknowledge and appreciate the complexity that fosters life.

While ecological consequences of a nature–culture divide are generally easily recognized, ways in which this historical ontological divide transcends the physical relationship between humans and nature and underpins a domination worldview are less clearly apparent. Western cultural prejudices and injustices carry this problematic ecological orientation into the social sphere. Environmental justice integrates class, race, gender, environment, and social justice concerns, argues Julian Agyeman (Agyeman & Evans, 2004, p. 155). Looking into policy matters, he finds that the Commonwealth of Massachusetts in 2002 provides the

> principle that all people have a right to be protected from environmental pollution and to live in and enjoy a clean and healthful environment. Environmental justice is the equal protection and meaningful involvement of all people with respect to the development, implementation, and enforcement of environmental laws, regulations, and policies and the equitable distribution of environmental benefits.
>
> *(2004, p. 156)*

Civil rights leader, Martin Luther King (1967), captures the social dimension of interconnectedness, in a speech:

> In a real sense all life is inter-related. All persons are caught in an inescapable network of mutuality, tied in a single garment of destiny. Whatever affects one directly affects all indirectly. I can never be what I ought to be until you are what you ought to be, and you can never be what you ought to be until I am what I ought to be. This is the inter-related structure of reality.

A contemporary example of environmental injustice is the emergence of carbon markets that promise abundant profits for Western businesses while imposing forest use restrictions on indigenous people. This is a current example of how the human "needs" of the West are privileged above the ecological and cultural needs of indigenous people, who are closely associated with nature. Rather than reducing carbon emissions in the polluting industrial nations, the suggestion is made (in UN climate talks) that indigenous people—who have contributed little to global emissions and are paradoxically most vulnerable to climate changes— should restrain from cutting forests so trees may soak up atmospheric carbon. Yet indigenous people interact with trees as more than mere carbon sinks: trees provide material for building, clothing, fiber crafts, medicine, animal fodder, fuel wood, and sometimes even food. And, as the Chipko movement has demonstrated, tribal women in India have a spiritual relationship with particular trees that are fundamental parts of their community (Bhatt, 1990). Nobel prize winner and founder of the African Green Belt movement, Wangaari Maathai (2006), equally expresses the centrality of the physical environment to the sense of self and community when she proclaims, "I am as much a child of my native soil as I am of my mother and father." A failure to recognize these types of ecological, cultural, economic, and spiritual aspects of living systems as legitimate underscores the perpetuation of discrimination based in part upon a problematic ontological division of nature from culture.

Ecological Perspective

There are many ways to understand interconnectedness and ecological thinking. Interconnectedness has been eloquently articulated in scientific terms such as the Gaia hypothesis (Lovelock, 2000), quantum theory (Capra, 1975), and principles of ecology (Carson, 1956; Leopold, 1949); it also frames diverse spiritual understandings of the human place in nature (Kumar, 2002). The Native American phrase "all my relations" conveys a deep sense of interdependence. Gregory A. Cajete (2005) explains the educational relevance of this relationship:

> *Mitakuye Oyasin* (we are all related) is a Lakota phrase that captures an essence of tribal education because it reflects the understanding that our lives are truly and profoundly connected to other people and the physical world ... Education in this context becomes education for life's sake. Indigenous education is at its very essence learning about life through participation and relationship to community, including not only people but plants, animals, and the whole of nature.
>
> *(p. 70)*

Traditional American Indian education is built upon an understanding of the depth of relationships. Among the elemental points of this education described by Cajete (2005) are the universal traits of integration and interconnectedness, the principle of reciprocity between humans and all living things, notion of cycles within cycles, and value of learning through participating in and honoring relationships (pp. 70–71).

A view of the world as an interconnected existence may be expressed alternately in scientific or spiritual terms, but in sum, it is a simple fact, whether or not we are aware of it. To become ever more present to the mysteries and nuances of living relationships is to engage in a deep form of relational—or ecological—thinking. Ecological thinking refers to more than simply thinking about ecosystems; rather, it is a way of thinking which fundamentally refigures mechanistic theories of knowledge and in so doing challenges the hegemony of dominant worldviews that perpetuate disconnected mechanistic perspectives (Code, 2006).

We recognize a host of intellectual ancestors who have shined the light of consciousness on the interconnectedness of life and living things, including Chief Seattle, John Muir, Aldo Leopold, Rachel Carson, Vandana Shiva, Satish Kumar, and Fritjof Capra. As attributed to Chief Seattle, "Man did not weave the web of life; he is but one strand. What man does to the web of life, he does to himself." Aldo Leopold (1949) similarly recognizing the interconnectedness of all living things, asks of us to change the human role from one of conqueror of the land community to citizen in it. Leopold reminds us that we have an inherent interconnection to the earth and living systems, though it is increasingly obscured by industrial culture. These simple insights set a foundation for the environmental movement of the twentieth century. Rachel Carson (1962) further activates an emerging ecological awareness in *Silent Spring* by demonstrating the interconnectedness of living things through the health and environmental consequences of DDT use in agricultural fields. In looking beyond immediate outcomes she traces the less visible causal relationships of chemicals in the soil and plant matter beyond the myopic time horizons (one season) of the pesticide industries.

A half-century after Carson's work, Vandana Shiva (1993, 2000, 2005, 2008) has renewed attention to the disconnected mechanistic view which today continues to perpetuate widespread chemical use in agriculture with special attention to the social and ecological fallout associated with the Green Revolution. As Shiva (2005) observes "while we are rooted locally we are also connected to the world as a whole" (p. 5). Thus local actions such as application of fertilizer or pesticide must be understood within the larger context of an interconnected ecology; likewise it is an egregious error to view food crops outside of their integral cultural and ecological context. Corporations such as Monsanto fail to recognize both the interconnected impact of chemical use upon soil, water, or human health systems and the cultural impact of disintegrating balanced subsistence agricultural practices. Transitioning from polycultural subsistence farming to chemical dependent monocultural cropping thus disrupts social and cultural cycles in the same moment as it introduces literal poisons into the soil and water. Traditional farmers have selected diverse seed stock and make use of polycultural planting for a number of reasons, such as: ensuring crop viability in variable growing conditions, timing harvest to provide seasonal sustenance, co-producing of fodder, fiber, or mulch material, and valuing local cultural traditions and ceremonies. The imposition of genetically modified organisms or "miracle seed" crops dependent upon expensive and dangerous chemical inputs disrupts this finely tuned social, cultural, and ecological order, trading nuanced sustainable interdependencies among human and biotic communities for exploitative and destructive dependencies upon transnational corporations and the fickle global economy. Of final insult is the implication that the actions described above are in the best interests of indigenous people, defended in the name of "human rights" or justice. To this, in direct contradiction of the claim that chemical dependent monocultural cropping is necessary for human sustainability, Shiva (2005) explains that "restraint in resource use and living within nature's limits are preconditions for social justice" (p. 50). Moreover, Shiva (2008) reminds us that "dignity is an experience and consequence of self-organization and sovereignty, of sufficiency and satisfaction" (p. 45).

For Satish Kumar (2002), care of soil is interrelated to care of soul and society. This integrated understanding indicates that while each component may be cared for individually, separation is never really possible. For example, as one manures fields to return nutrients to the soil, the soul is nourished through participation in the life-giving order. In turn, society is nourished as healthy soil is prepared to create sustaining food. We cannot disentangle one from another, and do well to treat the trinity as a unit not to be disaggregated. Fritjof Capra (2002) indicates that one of the most profound lessons we learn from study of natural systems is that sustainable communities interact and develop intricate interdependencies with neighboring communities. Capra (2002) explains in *The Hidden Connections* that symbiotic relationships are of relatively greater importance in the evolutionary process than species competition, a revelation that confounds conventional understandings of neo-Darwinian evolution. This ecological insight is of great importance when we realize that the majority of social systems in the West have been designed to mirror understood biological reality, with competition as a central organizing concept. The value of interdependence and symbiosis in co-evolution has potentially transformative social consequences.

At present an ecological introduction to interconnectedness as a basic fact of life remains woefully absent from modern educational systems that educate as if consequences of destructive action do not exist (Bowers, 2000; Esteva & Prakash, 1998; Orr, 1992). While students learn skills for competition in the global economy, are they aware that the market exchange of ideas and capital ignores the interconnectedness of all living things? That the flow of information made possible by technologies such as the internet ignores the continuation of healthy living relationships between and among human cultural and natural biotic systems? Are students, who are unaware of the food–body–soil connection and lack the understanding of how and where food is grown, prepared for real life? In many cases the sad truth is that education as schooling is implicated in preparing a generation ill-equipped to understand the intricate connections of the web of life. The elaborate industrial abuse of land detailed and lamented by Muir, Leopold, Carson, and Shiva, among others, may be related with education. As David Orr (1992) has wryly noted, the social and ecological degradation of the twentieth-century is not the work of ignorance but of highly educated individuals.

Pedagogical Implications

With a goal of bringing life to schools and schools to life, our proposal of living soil as a relevant ecological metaphor brings a relational state of being to the center of educational organization and pedagogy. We insert living soil into an educational discourse presently guided by disconnected mechanistic and industrial frameworks with a view toward shifting the conversation toward a simple enrichment of life and living systems. Rather than connecting education with economic advancement as is done routinely in the present mechanistic orientation, we suggest that education enhances the vitality of social and ecological systems. According to Nel Noddings (1992), "Kids learn in communion. They listen to people who matter to them and to whom they matter" (p. 36). In a similar vein, their passion and interest can be developed beyond math and language arts that have become the focus of classroom performance under NCLB. In order to achieve meaningful relationships through care, Noddings has several key structural suggestions that focus around continuity of the students' educational experience. Specifically, she proposes that schools should have continuity of purpose, continuity of place, continuity of people, and continuity of curriculum (Noddings, 1992, p. 72). She explains that continuity of purpose requires that a school, whose primary goal ought to be to cultivate moral and caring students, clearly and explicitly spends time everyday developing these qualities in its students. For continuity of place, she advocates that students stay in the same school, preferably a smaller school, for at least three years. This helps to foster a sense of community and belonging for the student. Related to this is continuity of people, which requires that teachers and students maintain their relationships for several years. Finally, continuity of curriculum provides a variety of equally valued areas of study that are all tied to a central curriculum of care (pp. 72–73). Extending her argument for care in schools, we believe that the learning gardens not only teach care but the gratification that comes from seeing things grow provides incentives to students to stay engaged.

In keeping with Howard Gardner's (1999) theory of intelligence, Noddings asserts that education should expand its focus from simply linguistic and logical intelligence to also include spatial, bodily kinesthetic, musical, interpersonal and intrapersonal intelligences. Later, Gardner (1999) also included naturalist intelligence. As Noddings (1992) writes, "it is clear that any education that values and builds on only a part of human capacity cannot be the best education for all" (p. 31). For her, curriculum that values all forms and capacities for learning are to be valued since children's multiple talents and unique abilities are cherished. Noddings' central premise that schools should be focused on developing students who are healthy, competent, and moral people is at odds with a national agenda of education disconnected from life.

With life as a focus it is possible to connect compounding concerns of nature deficit disorder, childhood obesity, food security, ecological degradation, and the unraveling of social safety nets. Learning gardens are one example of a whole systems solution that can be implemented in every school across the country directly on the school grounds outside the building: through connecting learning with eating, growing with harvesting, caring with consuming, and so forth, learning gardens serve as a nexus of interconnectedness. More specifically, the living soil of learning gardens is itself a vibrant micro-ecosystem of interdependencies that we can learn from as a guide for organizing sustainable social systems. As a dynamic metaphorical guide, living soil brings attention to relationship as a central feature of education: relationship among students, teachers, subject matter, school and community, school and nature, etc. Most significantly, living soil casts such interrelationships in a life-giving capacity.

Examples from Learning Gardens

It is imperative to consider that humans are vital members of the soil community (Hyams, 1976; Kumar, 2002; Shiva, 2005, 2008). We not only harvest from the soil but also contribute to its nourishment or depravity through our actions. Human life is interconnected with soil in an endless dance—and the key is to recognize this hidden connection. When we turn fava bean plants into the soil with children, a question usually comes up regarding why we are killing plants and covering them with soil. "Isn't the whole idea of learning gardens to grow living plants?" children often ask. A basic overview of the nitrogen cycle and visual exploration of the root hairs helps to explain the purpose of our actions. What is more striking than children's innocent questions is the general ignorance of the nitrogen cycle among the general public. Before synthetic nitrogen was developed (ironically as a peacetime application of chemical weapons used in World War II), soil fertility was maintained exclusively through crop rotation and the incorporation of animal manure, two practices which had the related effect of limiting agricultural scale. Learning gardens can be one venue for re-teaching the nitrogen cycle as a basic principle of living soil, and in so doing, highlighting the primacy of interconnection that reflects a basic principle of ecological systems: "to be is to be related, for relationship is the essence of existence" (Swimme & Berry, 1992, p. 77). Through broad-scale planting of fava beans, white clover, or other leguminous cover crops in the living soil of school learning gardens, students participate in gathering nitrogen from the atmosphere through hundreds of individual plants, which, like river tributaries or tree branches, distribute the harvested nitrogen through their subterranean root systems, as water is distributed through river deltas or transpiration through tree leaves. This activity is a first step in re-learning a pattern of literacy and understanding the network patterns of organization that sustain all living systems. The mycrorrhizal fungi networks that are present below the ground and in the soil originate from plants and establish mutualistic symbiotic relationships that play a critical role in soil fertility and provide nutrition to plants. The mycrorrhizal networks are fascinating to children.

Nitrogen Fixation: Learning Symbiosis

Engagement with living soil in school learning gardens is one practical way to introduce students to vibrant dynamic ecosystems, complex networks of material and symbolic communication, and the very idea of interconnectedness. Soil is composed of a complex web of relationships among micro-organisms, small animals, living and dying plants, tree roots, and mycrorrhizal bacteria, which together regulate the system in dynamic stability. A myriad number of relationships bind these organisms together. As described earlier, leguminous plants such as those in the bean and pea family host symbiotic nitrogen-fixing bacteria in their root nodules; atmospheric nitrogen is thereby captured, and made available for plant root uptake. This complex network of specific types of plants, soil bacteria and micro-organisms is together responsible for maintaining soil fertility in the absence of nitrification through lightning strike or chemical additions, and is the main method of organic agriculture in the form of rotating nitrogen-fixing cover crops such as beans, crimson clover, buckwheat, and alfalfa.

Planting a Three Sisters Garden

In diversified ecosystems, interconnection among parts encourages resiliency as a response to unexpected changes which are the nature of complex systems (Capra, 1996). Diversified networks of communication ensure that the "message" will be able to be delivered against many obstacles; diversified polycultural crop plantings ensure harvest in variable growing conditions. This is the reason for redundancy in living systems whether they are internal organs or ecosystems. The observation of redundant diverse interconnected systems inspires whole-system design methodologies such as permaculture. Bill Mollison (1988) indicates that inter-active diversity ensures stability of a system: "it is not the number of diverse things within a design that leads to stability; it is the number of beneficial *connections* between these components" (p. 32, emphasis added). Polycultural plantings, often referred to in permaculture as "guilds," intentionally combine one or more symbiotic plants in an attempt to mimic the interconnectedness of natural systems (Hemenway, 2000).

One well-known example of polycultural planting is the Three Sisters Garden. The Three Sisters is a Native American interplanting production system that was traditionally practiced throughout a broad range from the eastern woodlands to parts of Mexico (Gliessman, 1984; Sachs, 1996). Corn, beans, and squash are grown together in mounds of soil. This practice is associated with the Corn Mother deity, from whose body the first corn plant is believed to have risen (Sachs, 1996). The three plants are known as the Three Sisters because they demonstrate a symbiotic relationship: the beans climb the corn, which acts as a natural trellis; the squash spreads over the bare ground, keeping weeds down. This polyculture is a micro agro-ecosystem, and is representative of effective diversified interconnectedness. When European colonialists reached North America, they encountered corn for the first time, which native women had been breeding for nearly 10,000 years. Native people shared corn with their visitors, and introduced them to an agro-ecological approach that had been developed and perfected over several thousands of years: the Three Sister guild. Today, planting Three Sisters in the learning gardens revives ancient knowledge. The health and vibrancy of the Three Sisters relies upon open feedback loops and diversified interconnected living systems. We describe this example in school learning gardens as a practical way to explore interconnected diverse cultural and agroecological methods.

The Three Sisters (corn, bean, and squash-or-pumpkin) garden has been a popular garden activity over the years at each of the school gardens where there has been enough land to have squash trails. Students often read Native American traditions and stories associated with the Three Sisters. They also learn what the relationship is between these three crops and the science of companion planting. Since this is an ancient method of gardening using an intercropping system, students begin to observe and understand how mounds of soil support different crops in the same area and also how the crops relate to one another symbiotically. The corn, being the oldest sister, stands in the center, tall and strong. The squash, the next sister, grows over the mound at the base of the corn and protects the older sister from weeds along with shading the soil and keeping her base moist and cool with her own large leaves. Finally, the third sister, the bean, climbs through the squash and up the corn's stalk binding both the squash and the corn together as she reaches the sun. Sister bean also keeps the soil fertile by fixing nitrogen in her roots' nodules.

In first and second grade; blended classes at Sunnyside Environmental School in Portland, Oregon, students worked on projects related to the Three Sisters and developed posters for the Harvest Fair in October (Figures 9.1 and 9.2).

FIGURE 9.1 Students' research on beans

FIGURE 9.2 Students' research on pumpkins

Students engaged the visitors in interactive posters:

- How many kinds of beans are there? (Three—string, string-less, and snap)
- How many calories are there in a cup of green beans? (Only 44)
- How can you tell how many seeds are there in a pumpkin? (By counting the number of lines on the outside).

They had clearly learned to observe with care as they "researched" the life cycle of corn, bean, and pumpkins/squashes.

Students marvel at the science, the art, and the beauty of the Three Sisters. At Lane Middle School in Portland, Oregon, with a culturally diverse student body, vibrant stories emerge during a unit on corn celebrated as part of the Harvest of the Month curriculum. This is captured by the teacher in her reflective journal:

> Santiago opens up to the entire class … and shares a story about him and his grandpa. "My grandpa grows corn in Mexico. Every summer he puts me to work in the fields. It is hot! He says it is good for me … good for me to see how hot it is and work with the corn. He also has chickens. We eat the corn too." Rosa tells us about how her mom grinds corn on a grinding stone. The *metate* I brought in serves as a focal point for the students. Rosa approaches the *metate* and shows us how her mom works. "She mixes

the *maza* with water to make tortillas. We use the corn in mole also." This creates a stir in the classroom.

(Anderson, 2009, p. 75)

The ways in which students actively share their family stories and participate in learning highlight the connections between food, culture, and academic engagement. Corn is now more than a subject to be passively learned about. It is a place of stimulating connection that draws students into the conversation.

A picture of *elote*, a Mexican treat of corn on the cob, covered with mayo and sprinkled with chili, elicits many "yums" and "Oh, I love that!" A shy girl in the back slowly raises her hand and looks down. "My grandma lives in Africa and she eats a lot of corn."

(Anderson, 2009, p. 76)

Students also often wondered about how plants could be in such relationship where they help one another. Creating the soil mounds at the base of the corn so that her stalk can stand strong and tall was a puzzle for them. They also loved planting the beans an inch into the mounds and watering and watching them grow over the summer, especially in schools where there are summer programs. Moreover, students are surprised and puzzled by the beauty of the bean tendrils and the vine wrapping herself on the corn. The squash (or sometimes pumpkin that is planted instead) is especially inviting since it is usually ready by the time students settle in after the new school year begins. And the seeds, no matter when they are harvested, are particularly intriguing for students. This interplanting technique is applicable in many school gardens, even where land area is limited, as companion planting works to optimize space. From the Three Sisters Garden we observe the vital role interconnectedness plays in at least three areas: culture, ecology, and learning. The traditional planting method honors cultural understandings of integration and values relationship between people and all living things—two principles of indigenous education described by Cajete (2005). The biological and ecological relationships encoded in cultural mythology ensure sufficient soil nutrition through supporting symbiotic networks among mycrorrhizal bacteria living in root nodules of legumes. Through planting corn, bean, and squash together, children honor both cultural and ecological health and nurture interconnections between these two realms of understanding. Finally, and most significantly, Three Sisters Garden eloquently articulates the central role of relationship in learning.

Closing the Loop: Learning Gardens, Living Soil, and Sustainability Education

The cold self-interested rational mind is believed to be able to guide modern society, or at least a market economy. In this view, what is left of our humanity?

Modern science has confirmed what spiritual practices from around the world have known for millennia: we are more than just our brains, and the mind is always embodied (Capra, 2002). Mind and matter (or body) therefore create an interconnected whole that is substantially more than the sum of its parts. Despite growing general awareness and acceptance of this reality, the elevation of mind over matter and the attendant discourse of disconnection are still influential today.

The soil community produces no waste, recycles all products fully, and fundamentally promotes life. This final aspect is of significance in terms of our extension of living soil as a metaphorical construct for guiding education with a view to bring life to schools and schools to life. The living soil, upon which all life depends and which itself is teeming with life and aliveness, thus provides a potent example of the life-giving order of interconnectedness as evident in a Three Sisters Garden and the symbiotic nitrogen-fixing relationships. In the study of living soil and other ecological systems, we find a mode of interrelationship that actively contributes to life. The learning gardens express at once an education that is culturally relevant, ecologically appropriate, and pedagogically sound—a step toward sustainability.

10

AWAKENING THE SENSES

[By] re-awakening our senses and intentionally honoring subjective experience [we] return to our essential, animal selves, the selves that evolved in relation to the non-human natural world ... our sensory systems are exquisitely evolved channels for translating between 'in-here' and 'out-there.'

(Sewell, 1995, p. 203)

It is important to listen to nature because nature is so beautiful to look at! Why shouldn't it be just as beautiful to listen to? Nature can be so beautiful to all of your senses if you just take the time.

(Student, Learning Gardens)

With a goal toward bringing life to schools, we can advance an ecological per-spective through awakening and using the senses, including sight, sound, smell, taste, and touch—not just the eyes and ears. Awakening refers to enlivening the hitherto dormant senses such as smell, taste, and touch, and sharpening senses such as sight and hearing that are conventionally used in schools. Opening up and accessing the full range of sensory capacity can have a transformative impact in bringing life to school; learning gardens are naturally rich sites for sensory engage-ment, as they are filled with fragrant blossoms, thorny and prickly vines, delicious fruit, rustling leaves, and colorful flowers. In this chapter we elaborate a number of contributions the senses make to bringing life to an ecological education inspired by living soil as central metaphor.

Sensory Experiences with Soil and Learning Gardens

One of the rich lessons we glean from learning gardens is the value of slow and protracted observation over time for learning: we have seen children and youth engaged with the living soil and vibrant life of learning gardens listen deeply to silence, observe plants with great detail, touch worms with uninhibited glee, and smell herbs in dreamy repose.

The ecological and pedagogical implications are substantial, as such sensory and experiential learning is often effective over long periods of time, thus contributing to academic achievement and positive environmental responsibility. Bringing life to schools through awakening the senses unblocks sensory feedback loops, and encourages creative rethinking of guiding metaphors. Nurturing all the senses in the physical reality of learning gardens brings life to schools in the following ways: invite presence; support reminiscence and memory; ground otherwise abstract learning; and reinforce the idea of interconnectedness. Below we briefly articulate each of these contributions.

First, sensory awareness invites us into the present moment, the here and the now, which is the only place and time in which we can effectively act, feel, communicate, teach, or learn (Abram, 1996; Rosenberg, 2003; Tolle, 1999). Within the modern Western context, it is common to think and act from the mind alone without regard for the role of the body or of the embodied senses. Our senses live in the physical body. For instance, it is possible to taste fresh fruit here and now. While it is possible to remember how a fruit might have tasted in the past, or predict how a fruit may taste in the future, we can only really taste, feel, smell, see, and hear fruit in the present moment. The same can be said of teaching and learning. While it is possible to have an abstract idea about the future or the past, this kind of knowing is divorced from the now. Awakening sensory capacity invites a return to the present moment and encourages engagement with life. Second, because memories are associated with bodily senses, not just with the mind, the taste of certain fruits may bring back memories of childhood or family histories. The flavors do more than stimulate and arouse specific taste buds; they create physical connections that sustain personal memories and generate spontaneous sharing of stories with potential to create community. Meaning-making can be deepened through sharpening the role of the senses in learning.

Third, engaging the full range of our sensory capacity helps to center awareness, and grounds abstract concepts within physical reality. This aspect ties together the previous six pedagogical principles of sustainability education discussed in earlier chapters. Our senses are what allow us to make meaning of curious experiences in place. Though abstract ideas can describe general concepts such as "sense of place," grounded individual intimate presence is required to actually make sense of place; curiosity and wonder originate at least as much from sensory stimulation as from intellectual awareness. Awakening the senses encourages perception of less obvious natural rhythms and scales, and invites deeper respect for and valuing of the diversities that abound in an interconnected world. Practical experience awakens the physical senses in the most literal way. Finally, sharpening the senses reinforces in a bodily way the theme of interconnections that is characteristic of all living things. All of these themes are interrelated, as the physical body is naturally here and now, it is not abstract, and it forms interdependent relationship with mind and nature.

The full range of faculties complements knowing through the eyes and ears and integrates diverse components of lived bodily experience: our senses take us beyond intellectual understanding, opening a door that connects the living world inside to the living world outside. What escapes conventional "knowing" can be understood experientially through connecting with a complete palette of senses. For example, the rustle of leaves on a cold autumn night, the smell of coming rain, and the feel of a shovel parting soft earth are the types of sensory experiences that connect us with the living world and communicate critical information that cannot be codified or reduced to the intellect. Likewise, our physical bodies regularly provide sensory information regarding our state of being through signs such as a pit in the stomach, an itch at the back of the neck, or tightness in the knee. Tuning into our vibrant sensory capacity brings us into the present moment; listening, smelling, tasting, touching, seeing, and feeling are all activities that can only be done right here, right now.

Our senses alert us to the existence of layered relationships and multidirectional connections with all living phenomena. Indeed, even our sensory faculties themselves act in dynamic ecological interactions with one another. For David Abram (1996), one cannot help but perceive interconnections if one lives in the present moment and attends to information absorbed through all sensory pathways:

A genuinely ecological approach does not work to attain a mentally envisioned future, but strives to enter, ever more deeply, into the sensorial present. It strives to become ever more awake to the other lives, the other forms of sentience and sensibility that surround us in the open field of the present moment.

(p. 272)

In modern Western culture, all too often we close down this vital stream of information and rely on our intellect and mind alone for understanding. In much the same manner that modern industrial culture ignores the role of relationship in maintenance of healthy ecological systems, a privileging of the mind, eyes, and ears overlooks the relationship between these and all other bodily ways of knowing. Our rich sensory capacity is a fundamental aspect of what makes us human and connects us with the more-than-human world. This dynamic energy is blocked by lifeless mechanistic metaphors and industrial models of education. Awakening the senses is key to engaging with the life-giving abundance characteristic of the natural world. Learning gardens on school grounds are accessible places for engaging and sharpening the senses. By becoming present to our senses, as Abram describes above, we embrace our humanity as an ecology of experience.

Ecological Perspective

Moving back and forth between the metaphoric and material realms, living soil stimulates the full range of sensory capacity including sight, sound, smell, touch, and taste. Work in learning gardens brings children into practice using all their senses. We can encourage possibility for connection by transitioning toward educational models rooted in ecological phenomena such as living soil.

Farmer David Masumoto's (2003) *Four Seasons in Five Senses* captures the sensuality of growing, harvesting, and eating peaches not as an absentee farmer for whom peach might be a mere commodity but rather as someone with an intimate sense of the texture of soil, the feel for water and weather patterns, and the intergenerational connection to place. He revels in the tastes, textures, and aromas of his delectable fruit. As he explains:

> My peaches embody the uniquely brewed taste of a small family farm. I can't grow them in large volume, nor do I want to. I work side by side with family members, I grow my fruits organically, and the process is intended to be complex because I work with unpredictable mother nature.
>
> *(Masumoto, 2003, p. 5)*

His analysis of fast-food farming is a reminder of just how much "flavor" is lost when the normal rhythms of the peach's life cycle are disrupted and hastened through synthetic fertilizers or chemicals designed to stimulate growth. When peaches, like all food, are commodified and numbered for computers to decipher as we see in the computerized check-out stands in grocery stores, there is no telling one peach's individuality from another.

The sensual story of peaches has much to do with a deep understanding of place, work, farming, and the cycles of birth, growth, life, and decay, itself. Slow-food for a fast-food nation serves as an important metaphor for why sensitizing the senses is critical for our children.

Each one of us must eat to live; but there is a difference between buying packaged food obtained through fast-food drive-ins and savoring of food through the aroma of cooking or picking of a fruit directly from a tree, feeling its shape and texture, touching it, smelling it, and biting into it. With luck, the juice from the fruit would drip down the chin and roll down the hand, further awakening our bodily senses. The plucking of a ripe fruit often lets loose a deep aroma that titillates the bodily enzymes.

The multitude of smells in a garden are partially due to subtle differences in fragrance of flowers and fruits, easy to be recognized only if we pause and smell them: plum blossoms smell different than apple blossoms or mango blossoms. And, there are seasonal variations of smells based on who is rooted in a garden. Besides, no two plants necessarily produce flowers or leaves or fruits that are identical. In fact, even the same plant does not produce identical leaves or flowers. There is tremendous variation within a plant. In so far as smell is concerned, a rose serves as a good example. Given its tapestry of petals rich in color, not only are we awakened by our sight to the subtle differences in the myriad shades of red and pink, our olfactory sensibilities are also awakened to the myriad and subtle differences in the aroma and perfume sometimes meant to intoxicate. The significance of all this is that scents and memories are linked intricately: scents bring memories of life-stories—of places and communities—in fact, we sometimes long for those smells even as we re-live those memories. Our eyes and noses are inquisitive as they are meant to be.

What sustains our bodies isn't simply "calories" that we measure and are listed on containers and packages. No arithmetic figure can compare with the figurative arousal of enzymes resulting from the anticipation the body feels for food to be relished. For Masumoto, for example, interaction with his peaches serves as honored ritual, and in fact is viewed as reciprocity of relationship such that human care is essential to his peaches for them to provide joy and nourishment. The sun, the rain, the rich soil, and the careful pruning are equally significant in the final outcome.

A garden can also teach us about sound. While we routinely enjoy the smells and tastes of a garden, we can train our senses to measure sounds as well. Often forgotten in appreciation of plants and trees in school learning gardens is the contribution of pleasant sounds, such as the rustling of leaves, sound of wind, and chirping of birds. Plant borders around schools also diminish noise from nearby roads that may distract learning. When we walk through learning gardens, the plants and trees seem to speak as they sway to the rhythm of the wind. Students engaged in quiet observation often find they are able to hear more and more layers of activity in the gardens. Masumoto explains his observation of the sound of peach:

> I know of two types of ripe peach sounds. One is the noise of eating a peach: the breaking of the skin as teeth sink into the flesh, the sucking of juices out of the meat, the first noisy chews with mouth open as the nectars wash our taste buds, and the smacking of lips and tongue.
>
> *(p. 97)*

Touch is also a part of life and how we know the world. Soft and hard, gentle and firm, are lessons babies learn at a very early age. The sensations of touch arise by the activation of sensory receptors located in the skin that are responsive to mechanical stimuli. However, the tips of our fingers are now so used to the hard keyboard that we are losing ways in which skins communicate. Gardens are rich for tactile literacy. Students engaged with living soil in learning gardens are able to use their senses when using tools. As Masumoto (2003) explains:

With my shovel I can read landscapes, feel the subtle nuances of soils, assess the moisture of a field. The tool can quickly dispose of a weed, yet with the intention of returning, for there will always be more weeds. Shovels force long-term thinking and imply that I will be around for a while.

(pp. 92–93)

A garden provides sensual pleasure through stimulation and diversity.

Pedagogical Implications

In schooling, the full range of senses is narrowed to only sight and sound; other senses are marginalized as unimportant or viewed as distractions. Visual and auditory senses are heavily stimulated by reading, writing, speaking, and listening activities in schools, and are commonly mediated through use of technology, thus creating vicarious experience of phenomena. Use of videos, computer games, PowerPoint, and "smart boards" as educational "tools" are examples of visual technology. While each of the above examples has their place in education, a reliance upon technology as educational panacea narrows the range of life experience to which children are exposed. An Apple McIntosh is not an edible apple and a byte of data has no nutritional value. Likewise, attention focused on a Windows' screen instead of through a window screen only allows children to connect with life through replicated images rather than through authentic observation. Reduced to binaries of 1s and 0s, it creates an illusion that the mysteries of life can be unlocked with the touch of a button. The quality of observation mediated by technology is markedly different than what is possible through direct experience. The same observation can be made of auditory information. Also forgotten are the complementary senses such as smell, touch, and taste, all of which can be affective avenues of expression, articulation, investigation, and understanding. As the senses are connected to life, the awakening of sensory awareness is a crucial step toward bringing schools to life. Moreover, awakening the senses is central to ecological thinking and animates the previously discussed pedagogical principles. In life, senses are the basis for all art forms. Along with learning to sing, paint, and draw, students' senses need to be awakened to see, to hear, to observe, to feel, to touch, and to move.

Today, it is not uncommon for a majority of children to develop perceptions and understandings of the world mediated by the influence of technology. Electronic modes of communication highly stimulate mostly the eyes and to some degree also the ears; however, at what cost? Simply stimulating the eyes through reading text or watching a screen, ignores a fundamental aspect of what makes us human—our rich sensory capacities. Together, touch, taste, sight, smell, and hearing, assist us with embodied learning, developing in us sensual awareness, often unspoken. Our sensory capacity supports presence in the moment.

While we attempt to codify the diversity and complexity we encounter in the world of nature, once we become aware of its magnificence, we have to acknowledge the role that our senses collectively play. According to David Abram (1996):

> By linguistically defining the surrounding world as a determinate set of objects, we cut our conscious, speaking selves off from the spontaneous life of our sensing bodies. Only by affirming the animatedness of perceived things do we allow our words to emerge directly from the depths of our ongoing reciprocity with the world.
>
> *(p. 56)*

Echoing Abrams, ecopsychologist Laura Sewell (1995) indicates that our sensory systems are exquisitely evolved channels for translating between "in-here" and "out-there" (p. 203). According to her, awakening and sharpening the senses involves intentionally honoring subjective experience and teaching children to "return to our essential, animal selves, the selves that evolved in relation to the non-human natural world" (Sewell, 1995, p. 203). Learning gardens on school grounds can provide a living laboratory for practicing sensory engagement.

Examples from Learning Gardens

Teaching children to attend with their senses is foundational to developing connection with life. As biologist Edward O. Wilson (1984) states, in the making of a naturalist citizen, it is "better to be an untutored savage for a while, not to know the names or anatomical details. Better to spend long stretches of time just searching and dreaming" (pp. 11–12). The modern world is so planned, safe, and predictable that our sensory awareness to life is shutting down. Instead, we have seen educators drawing upon the process of allowing individuals of all ages to find a "sit" spot or "sacred" spot in nature and to learn to be quiet, and to begin to listen and observe with care and attention (Young, 2001). This can last for ten minutes or an hour. But revisiting the same spot is the key since this encourages children and youth to sit quietly at a consistent place in nature and helps them learn to pay attention, to be mindful, to observe intently, to understand seasons, to marvel, to wonder, to listen with care, and to breathe. Our minds are conditioned to wandering. Instead, by directing focus on one place, a constant, and one sense at a time, students build sensory awareness through practice. This provides them a way to connect with the natural world. The sit spot is often called the sacred spot since it creates an awareness of wilderness that the modern hectic life and school life does not normally permit. As a doorway to a world of quiet, in some sense, the experience makes students more contemplative.

Sit Spots: What Senses Convey

Children and youth at Lewis Elementary School in Portland, Oregon, are encouraged to sit in the gardens with no particular goal in mind. With closed eyes, their sense of smell and touch are activated as they try to inventory the plants in the gardens on school grounds. They begin to learn the reciprocity occurring between themselves and the species around them. In one of the fourth grade classes, students are taught to be attentive and observant and to listen with all their senses. A student shares her thoughts on how this helps her:

> Why should we listen and observe around the garden? I think we should because if we talk it is distracting, and the only connection we can have with the plants is observing them, then we can feel at peace with nature. My secret plant that I chose is by the big pine tree. There are wild peas, mint and many other plants that I want to learn about.

Another student, for whom mint is her secret plant, writes:

> Today while gardening we learned that mint can calm, settle stomachs, and help with headaches. I didn't know that just one plant could do so many things and still taste so good! Mint reminded me of when I was younger and I would go to my grandparents' house. My grandma loved to garden and I loved to help her … I would sit in her patch of mint with my eyes closed, smelling them and munching one. That was probably the best thing to do on a hot summer day.

This student is sensitive to the larger context within which her senses are revived and memories are aroused. She further elaborates:

> It is important to listen to nature … plants give us so much, and we give them back so little. Yes, we water them and feed them, but they can do this all on their own. So, if you listen, I mean *really listen* to nature, you can understand what they want. Try it! I can't do it that well, even though I try (emphasis added).

Plenty of experiential opportunities to engage the senses arise in learning gardens on school grounds. There is the depth of sensory opportunity and engagement ranging from the soil, to flora and fauna, all the way to the zenith, including the horizon and the sky. The voice of a seventh grade student at Sunnyside Environmental School captures this best in the following writing:

Brookside

As the river runs past
I can hear the birds sing from in the trees

And hear the branches rustle
I see the small critters crawling through the brush
The rock under me is cold and ruff
I smell the wet dirt
And hear the cars in the distance
It is very peaceful here

In Lewis Elementary School, students like to taste a variety of vegetables from the gardens they have planted. Their sensory revival is evident in the comment below:

Today we made pesto sauce. Instead of having an electric masher we mashed the pine nuts and garlic and basil with a mortar and pestle. I ground the pine nuts into a paste that smelled sort of like peanut butter … The pesto tasted leafy … I am going to plead with my mom to make pesto tonight.

As seen in the above expression, taste buds are activated as students sample the vegetables they have grown. Similarly, this is promoted in several school districts across the country, most notably by restaurateur Alice Waters at the Edible Schoolyard at Martin Luther King, Jr., School in Berkeley, California (Stone & Barlow, 2005). As many schools have eliminated kitchen cooking facilities in an effort to cut labor costs and maximize efficiency, pre-packaged food is standardized based on federally mandated caloric values, mass produced, delivered to schools, and sometimes reheated for lunch, thus reducing the sensory experience of eating to dull monotony. School cafeteria waste audits indicate much of the prepackaged food is discarded without even being opened. In order to overcome the homogenization of culinary experience and encourage healthy eating, schools are revitalizing kitchen cooking facilities in connection to school gardens. The aromas of cooking food engage children's senses in diverse ways, including smell, taste, sight, and tactile skills. The serving and eating of real cooked food creates a living ambience.

Cob Construction

Another sensory experience that is possible is cob construction. Several students from Lewis Elementary School in Portland, Oregon, were involved in making a cob bench for their learning garden at the school as part of their outdoor classroom construction project. The students worked with several partners: the Village Building Convergence, a community non-profit that supports environmental and sustainability projects in Portland, a professional natural builder, and several parents, students, and college partners. As seen in Figure 10.1, students use their

FIGURE 10.1 Feeling earth

feet to mix mud, clay, sand, hay, and water, to prepare the building construction materials.

At the Lewis Outdoor Education Center, the cob bench serves as an outdoor classroom where students can sit and discuss garden projects with their teacher and garden coordinator. Students learn about water conservation, gardening, native plants, and composting. Included in the Outdoor Center are also raised vegetable beds, rain barrels, a greenhouse, and a revival of a decades-old orchard of apple trees. The outdoor classroom also has an eco-roof. The students learn about seed saving, herbs, organic gardening, and growing starts, along with water conservation techniques, food waste audit, and separating food waste to maximize composting and minimize dumping. One of the students, who was involved in building the cob bench, describes his experience thus:

> One day we made cob for the cob bench. We took off our shoes and we got a bucketful of sand and then clay. We dumped them on our tarp and we mixed them together with our feet. Then we added straw. Again, mixed them together with our feet. We added water then made little [clumps] of it.

Even after graduating from the school, students return to the school to their cob bench as they have good memories and are proud to have made it.

Edge Experience: Cracking Asphalt

We are aware that it is not easy for students who have not been exposed to nature to actually "know" through their senses. An experience that challenges the student's sense of the known and that calls on his senses to perceive just a little beyond what he already knows, is called an edge experience. These experiences cause us to see things in new ways, growing in perspective with each such event. It is stretching into the edge of the unfamiliar, not the familiar. Such edge experiences expose gaps in heretofore held static perceptions of an objective world. Edge experiences reshape the known world as fluid and ever-changing and call upon us to awaken all of our sensory faculties. In the words of poet Annie Dillard (1974):

> The gaps are the spirit's one home, the altitudes and latitudes so dazzlingly spare and clean that the spirit can discover itself for the first time like a once-blind man unbound. The gaps are the cliffs in the rock where you cower to see the back parts of god; they are the fissures between mountains and cells the wind lances through, the icy narrowing fiords splitting the cliffs of mystery. Go up into the gaps. If you can find them; they shift and vanish too. Stalk the gaps. Squeak into a gap in the soil, turn, and unlock—more than a maple—a universe.
>
> *(p. 274)*

Education that awakens the senses invites students to explore the "gaps" Dillard describes through edge experiences. Applying such edge experiences to soil, which an urban student might familiarly see as something peeking through cracked asphalt, might entail having students move a pile of soil from a recently de-paved school parking lot. Figuring out how to break the asphalt into smaller pieces and separate out the soil underneath would be a challenging but rewarding task; we have seen students take great pride in converting their parking lots into community gardens. As one student observed: "Five months ago, we had grey here; hard grey surface. Now it is becoming green and so alive and beautiful."

Closing the Loop: Learning Gardens, Living Soil, and Sustainability Education

Traditionally, in the classroom, children mostly use auditory and visual senses. On the other hand, a garden is rich in opportunity to sharpen the sense of smell, touch, taste, and kinesthetic awareness, as we see manifested in the student expressions below:

I look forward to the day of sitting once more on a stone drawing the beauty of a garden unfolding.
(3rd grade student)

I like to draw pictures of the garden. The fragrance of the garden makes me feel happy. Being all alone in the garden is peaceful. Touching the smooth colored pencils makes me feel like a real artist.
(3rd grade student)

You feel it, hear it, touch it. Instead of looking at a book, you actually work in the garden and try to plant a plant.
(6th grade student)

Sensitized thus, students are suddenly able to incorporate information on many levels, as we have seen with hundreds of children and adolescents over the years. Students not only learn the names of flowers and the parts of flowers (which is typically done in a classroom), but now they can taste some of the edible flowers, watch bees pollinate the flowers, find several other insects in reciprocal relationships, smell the flowers and leaves especially by crushing oil glands that some might carry, and differentiate the various smells. Picking, cooking, and eating fresh produce from learning gardens can be a stimulating sensory experience.

Students get in tune with seasonal variations, birth and life cycles, and changes within plants and in the garden over time. They are more likely to recall the intricacies of plant life a year later when immersed in such sensory experience, since awakened tactile and olfactory senses support profound memories. Engaging and sharpening the senses leads to the understanding of interconnectedness and Capra's (1996, 2002) notion of nested systems. It is possible for children to notice the nested systems in a garden: by using different senses and having edge experiences, children and youth understand the effect of solar energy by seeing more bees in the garden. They track daily and seasonal cycles by feeling the airflows on their skin. They discover the diversity of plants and animals including insects. They notice that humans can share space with birds and animals in a garden. They hear, smell, and move as part of the dynamic balance of life in the garden, and go home with firsthand experience of each ecological principle in support of sustainability. All of this has significance for education and ecology alike.

Practice Comes Alive from the Ground up

11

TEACHER, PRINCIPAL, AND SUPERINTENDENT PERSPECTIVES

In this chapter, we present perspectives and insights from three practitioners—a teacher, a principal, and a superintendent—who have intimate knowledge of and experience with gardens at their schools and/or districts. Our rationale for selecting educators at three "levels" of the school system is to show that at each level of the organization, there are educators who have a depth of understanding and knowledge of learning gardens as educational tools. Whether it is the classroom, the school, or the district, these individuals represent the voices of educators across the country where school gardens are designed, developed, and integrated with the curriculum. We have picked two school districts: Portland Public Schools with which author Williams has a 20-year history and where both authors have started and/or supported learning gardens in schools; and San Francisco Unified School District, which has a large percentage of schools with robust school gardens that have served as research sites for Williams since 2005. Both school districts are similar in size and have a large number of schools with learning gardens. They are both considered Title 1 school districts with over 40% of students on free and reduced school lunch and of non-white status.

We begin with Access Elementary School, a K–8 school on the Sabin campus in Portland Public School District, in north Portland in Oregon. In 2008, Sabin school's vacant lot was transformed into a garden. By using this story, we demonstrate how learning gardens can be started and integrated with school curriculum. We requested third grade teacher Deziré Clarke to share her experiences since she had involved her students in building the gardens. More notably, as a teacher, she designed a garden-based curriculum and activities for ongoing engagement of her students. We requested that she use her teacher's lens and perspective to explain the evolution of the garden and ways she and her students were engaged in the learning that occurred.

Next, we highlight a K–5 school, Lewis Elementary, in Portland, Oregon. Unlike Access Academy, this school is unique in that it has a 40-year history with outdoor learning that has seen waves of environmental interest. Yet, in the last decade, the school's garden program has been strengthened and serves as a site where visitors, both national and international, are inspired to see the vigorous parental and community-based support for the principal's vision of integrating their learning gardens into the broader ethic of sustainability that the school has embraced. Tim Lauer, the principal, shares his perspective as a school leader with a focus on partnerships which he views as critical to such ventures and school learning.

Finally, we present Superintendent Carlos Garcia's perspective on school gardens in San Francisco Unified School District and ways in which his district has created an infrastructure that continues to foster the building and integration of gardens in the San Francisco schools. Bonds passed in 2003 and 2006 included green schoolyards construction and the decade-old strong grass-roots support for gardens has been sustained.

We believe that the voices and insights of these practitioners will demonstrate in authentic ways how practice comes alive and will help move forward not only the creation of learning gardens on school grounds but also ways they are to be integrated to ensure that academic learning is promoted. Soil's living features can thrive in our schools when we value life-affirming language and practice at all levels—the classroom, the school, and the district.

Access (K–8) Academy at Sabin, Portland, Oregon

Sabin Community Native Garden and Sabin Edible Garden are the result of students, teachers, parents, and volunteers coming together from the Sabin neighborhood school community in Portland to bring vitality to the neighborhood and the school. Between 1990 and 2000, the enrollment at Sabin Elementary School (K–5) was declining. To avoid closure of the building, Access Academy (K–8) was started and housed in the building to serve gifted students in the Portland metropolitan area who performed at the 99th percentile level in nationally normed tests for aptitude and/or achievement in any or all of the following areas: language, math, or general intellect. The mission of Access is to develop a learning environment so that gifted children thrive socially, emotionally, and academically.

In May 2007, dozens of Sabin parents, teachers, staff, and neighbors laid the ground work to begin plans to convert a vacant lot at the school which had a bland expanse of unevenly growing grass into a garden. The north end of the school was selected since there was open space and sunlight and the area could be potentially fenced. The foundation was laid for the start of the Sabin Edible Garden with a view to making the space available for educating students at Sabin, Access Academy, and the Schools Uniting Neighborhood (SUN) schools program. SUN schools, a program for low-income students, operates both during the academic year and also the summer; Valerie Thompson was hired as an AmeriCorp SUN staff and she joined Madelyn Mickelberry-Morris, hired by the Sabin Parent Teacher Association (PTA). The PTA comprised committed parents who wanted to revive the neighborhood through the garden program and provided financial support and human capital for the Sabin gardens. Over time, Isabel LaCourse was hired as coordinator. Besides vegetable, wildlife, and native garden beds, attractive benches and pavers for stone paths were constructed and placed in garden areas where there is ongoing development of beds and art and educational activities.

The schoolyard was transformed from a barren lot into an edible garden that was designed in the shape of a star with separate raised garden beds and barrels. Fresh vegetable starts went in, as well as fruit trees and berry bushes. Students, parents, and other volunteers spread donated cedar chips around the pathways to suppress weeds, maintain moisture for raised beds, and give the garden a beautiful and well-maintained appearance. As the garden began to become more established, the PTA identified the need for a Sabin Garden Committee to be responsible for allocation of resources, funding for projects and materials, garden maintenance, curriculum development for classroom use of the garden, long-term goals and vision, and the distribution of food grown in edible gardens.

As interest in the Sabin gardens grew, more parents with gardening, carpentry and construction skills volunteered. A beautifully designed iron gate was installed. Two cob benches were built with sand and straw with student involvement. Garden art made by students added color and beauty. The clay art was put in by Access Run for the Arts funds in 2009 with guest artist Sara Ferguson. In addition, on the south and west side of the school grounds, the new Sabin Community Native Garden received approximately 300 native plants and had pathways redefined with fresh bark chips.

The neighborhood takes pride in a learning environment that connects children to nature. Students harvest food that grows in the gardens and they are often outdoors learning about wildlife and nature. When the garden is hibernating, students plant cover crops to protect and enrich the soil and harvest hardy winter edible crops on a regular basis.

We invited Deziré Clarke, a third grade teacher from Access Academy, to share how she designed the curriculum and integrated the Sabin gardens in her teaching. The next section captures her philosophy of teaching and elucidates ways she spent a year involving third graders in learning in and from the school garden.

Ethnobotany: A Year of Schoolyard Learning—Deziré Clarke, Third-Grade Teacher

I would like to address who I am as a teacher and how I plan for my classes before I can talk about the garden program, since that will help with providing the broad context for how my students were engaged in schoolyard learning. Much of what I teach my students is a deep appreciation of the natural world and a passion for lifelong learning. I have an open-ended approach to teaching, but I structure my curriculum in ways that pull all students into the conversation. I wanted to weave gardens into the third grade subjects that I teach; there is no particular book that can teach one how to do that, step-by-step, for everyone, because gardens are so season-specific and also schools are community-specific. But there are many school garden books that are inspiring and encourage us to try as teachers. The important thing for me was to be true to my beliefs regarding an integrated approach to curriculum across disciplines, but also to look for ways to involve my students in a hands-on practical approach to learning about gardening. It was about teaching across subject areas and the skill of gardening, since nature, I believe, is a teacher.

In 2007–2008, as a new third grade teacher at Access Academy, I wanted to help foster in my students a deeper appreciation of how the natural world sustains us and promotes the environmental and social well-being of our school community. To really have my students understand the importance of the local, I needed to actually look at our own school site, not just Portland, or the Sabin neighborhood. A school garden, I thought, would be excellent as a way to engage students to appreciate nature.

My hope after a summer of planning was to place garden bed frames with my students' help on the broken concrete outside my portable classroom for the potential of growing and learning by building a garden. I returned to the school at the start of the school year in 2008 eager to organize my classroom and create space for seeds and minds to explore and grow. I noticed the improvements that had occurred on the school grounds over the summer: graffiti had been painted over with a beautiful mural in its place, and the most astounding improvement was the schoolyard corner. I walked bemusedly over to this new space to find raised beds arranged in a sun burst pattern with fruit trees and wood fences at the borders. The school year had started and my main question was: how could my students and I get involved with this new plan for a garden?

Planning Curriculum to Include School Gardens

Instead of following the traditional disciplines, I asked myself, how can I weave the garden into the curriculum? First, I framed this within the broader context of "place-based" education because of how important it is to teach students about their own local community. Before weaving the garden into the curriculum, I was thinking of "community" and "local" as they related to "neighborhoods" and the city of "Portland." The garden at the school made me realize that I needed to ground my students in the community where the school was located. This is because ideas of "community" are tied with the concepts of place. So, I had to think about two things: first, how to get my students to become involved in building the gardens especially when we had concrete outside our portable classroom; and second, how to integrate the various subjects to make the garden more relevant.

In a traditional curriculum, the subjects are separated from one another, and, if schools can afford it, they might offer art and music. However, nature, the outdoors, community, service, and true life experiences are remotely connected to the curriculum. This involves a separate approach to subject areas, and value is placed on certain subject areas and not others when the curriculum is designed. I could not accept this model because I believe in a holistic, integrated curriculum.

In trying to get ready to build a garden at school, I asked myself, what connections and curricular themes would I employ to engage my students? I developed overarching guiding questions instinctively: Why is "local" important? What is the importance of a garden? How can we make a positive impact on our local environment and community through the school garden? I visited the library and local bookstores to find picture books and novel sets that would be relevant to the curricular topics for the year: Oregon history, economy, agriculture, and geography. Most importantly, my search included books that discussed issues and supported ideas for gardening with children. Picture books can provide an excellent lens for students to identify with deeper meanings and utilize higher-order thinking. In the past, I have used picture books to illustrate many complex issues concerning people and life experiences. Throughout my search, I found books concerning big topics like hunger, charity, social change, community, and activism. Beyond the book lists I began to compile, I inquired with librarians and storeowners about local responses currently in practice around my school community concerning these topics.

I began to look for ways in which the curriculum could now reflect the various themes from the subjects I had to cover. I began to develop an integrated curriculum. Whole-child learning is accomplished when subject areas overlap to create interconnected experience utilizing outside community and building a strong sense of classroom community. Creativity, imagination, critical thinking, problem solving, decision making, and physical, social and emotional skills are at the center of interconnected whole-child learning. This contrasts the distant location of creativity and imagination at the boundaries of the traditional curriculum model. Additionally, one can see how all subject areas are interrelated to one another and connected to the garden context, which itself is further embedded within the larger community, which includes family members and outside experts. Subject areas are like members of an ecosystem, coming together to create a whole that is greater than the sum of its parts. Finally, the boundaries of the integrated curriculum model are much more permeable than those in the traditional curriculum model, moving learning beyond the four school walls.

Being an elementary school teacher, I need to consider multiple subjects when planning my curriculum. This approach helps me keep in mind the whole child.

I always ask, what content and developmental needs will be addressed when approaching large curricular themes/concepts? The curriculum I build is current, integrated, and pertains to local needs. With this in mind, I explored ethnobotany since I was interested in teaching students not only about plants but also about culture, place, and community. Ethnobotany is the connection among people, plants, and places. So this served as an organizer of the curriculum for a whole year of schoolyard and garden learning.

Methods of Teaching

I used several approaches to teaching: Building Classroom Community; Science Talks; Community Mapping; Designing a Book of Students' Writing called *Seedkids* modeled after Fleishmann's (1999) *Seedfolks*; Technical Drawings; Making Tea with Herbs; Field Guides; Journaling; and Hands-on Gardening. I will explain these in the context of my teaching inside and outside the classroom.

Building Classroom Community

First, I had to build a classroom community. Why? My goal as an educator is to provide as many opportunities as possible for participatory education. Every year I begin my class by asking students: how do we recognize each other and value each other as we share this space? I explore with the students ritual and meaning-making by establishing norms. Meaningful time is set aside each day for the celebration of everyday relationships. Human connections can provide an antidote to the cultural pressures of individualism and consumption. Purposeful activities are planned where students recognize their role in creating a harmonious classroom and in turn building concepts of community. With my third graders, we collectively developed a *Classroom Constitution* that helped with establishing the norms for the new year that addressed caring, courage, friendship, flexibility, and so on. The students developed a classroom constitution, which they signed and we framed.

Seeing the connection beyond the immediate self and family toward one of interconnection outward, students could be engaged in higher-order thinking beyond the obvious. For example, as homework to involve families, each student was asked to complete a "family tree." The collection of family history resulted in strengthening family bonds by sharing stories about ancestry. These stories helped children recognize that they are connected to others beyond immediate family in their households. Another assignment that students completed was to make inquiries about their name. They asked their families, what does my name mean? Why did you choose to give me this name? What's in a name? In class we read texts that concern the importance of names, like the picture books: *Chrysanthemum* (Henkes, 1991), *Pass the Celery, Ellery* (Fisher, 2000), *Catalina*

Magdalena Hoopensteiner Wallendiner Hogan Logan Bogan was Her Name (Arnold, 2004), and *Four Boys Named Jordan* (Harper, 2004). We read poetry and other forms of literature that examined the concepts of names. This helped with the ethnobotany unit and how plants are named and what they might mean. My goal as a teacher is to find links that make for deeper understanding for my students.

Science Talks: Making Meaning Together

Science Talks are discussions about big questions in Science. They are appropriate for any grade level, but particularly useful for elementary students. Students often generate the questions discussed during a Science Talk as it provides space for them to collectively theorize, to build on each other's ideas, to work out inchoate thoughts, and to learn about scientific discourse. Most importantly, this approach to teaching and learning allows students to do exactly what scientists do: think, wonder, and talk about how things work, the origins of phenomena, and the essence of things. Science Talks are a valuable tool for working on the culture of the classroom as well as working on social construction of meaning. Children sit in a circle and direct their comments to one another, not to the teacher. My task during a Talk is to listen carefully and to follow student thinking. Doing this provides a window on student thinking that can help me figure out what students really know and what their misconceptions might be. This insight helps me as a teacher to provide hands-on activities and experiments to feed their curiosity with things that puzzle them.

In getting ready for gardening and planting, we conducted a Science Talk about classification and naming. Examples of questions posed include: "How do trees stand up?" "How do plants grow from bulbs?" For instance, taking it further, in the talk on how plants grow from bulbs, one student commented, "Okay, I know that plants need sunlight, water, and nutrients from the soil in order to grow, but I don't understand how the plant uses those things. How does sunlight help the plant grow? How does water help? How does the plant get the nutrients from the soil?"

A series of experiments were now planned. For one experiment, we took a stalk of celery and placed it in water with red dye. The students observed the celery slowly turning red as the vegetable absorbed the water. This observation inevitably generated more questions and theories. They asked questions such as, "How does the celery suck up the water?" They made comments such as, "You'd think with gravity that wouldn't happen." My approach to answering these questions was not to rush in and give them these answers, but to provide opportunities for students to search for their own answers. I provide and direct them to books and resources on the subject to empower learning.

In another instance, looking at titles on a seed packet, students wondered why there would be so many names for one plant. When this inquiry began, we decided to conduct a Science Talk for the questions: "What makes a plant, a plant?" Students needed to categorize plants so from this Talk a further question evolved and was discussed: "What makes a vegetable, a vegetable?" Through background knowledge, observation, and connection students were grappling with the notion of scientific classification. The school gardens provided ample opportunity for students to be puzzled by what they were experiencing with plants and wildlife.

From Seedfolks to "Seedkids"

In one of my summer bookstore visits, I had acquired a small but powerful book, *Seedfolks* by Paul Fleischmann (1999). This book is made up of several short, interesting, and easy-to-read biographies all intertwined to tell a beautiful story illustrating the formation of a community by means of an urban garden. Among other personal insights, each character gives the reader a snapshot of childhood or adulthood experiences as an American. The individuals are of various ages and come from different cultural backgrounds. The community garden brings these separate, sometimes lonely, individuals together as real neighbors. Each character contributes to help create and protect the once vacant plot of land that became an abundant garden by the end of the story. For the characters in the story, the garden brings out a reason to talk and to show kindness, helpfulness, happiness, friendliness, and understanding toward one another and others in their life. Stereotypes slowly disappear and are replaced with the shared wisdom of each other's worldly experiences. Neighbors harvest more than just fruits, flowers, and vegetables from their community garden. They harvest feelings of community and friendship.

Seedfolks appears to be a simple book because it is short, but in fact it touches on very basic human needs and desires, especially how easily they are left unmet and unfulfilled. There is hope in the simplicity of how easily that can all be changed. I used *Seedfolks* to educate my students to meet writing benchmarks including perspective taking, narrative writing, and learning about analogy. But, when I used this book, within the context of our school garden involvement, students gained more. From this powerful little book, students gained inspiration to transform their own schoolyard.

Using *Seedfolks*, students explored the characters/neighbors that live in an urban apartment building in Cleveland, Ohio, through the activity of creating "character webs" as they read. A character web is a graphic organizer that assists students in taking note of various events or topics surrounding a specific character in a story. Working individually and in partner groups making character webs of each character engaged students further in analysis. Reading the book aloud encouraged my students to create "juicy sentences" using metaphor and simile in their own writing. We picked apart Fleishmann's text to reveal the secrets of good writing. We noticed what we enjoyed as a reader and then tried it ourselves as writers. We noticed a trend in the construction and format of this novel's writing. The outcome was a book (Clarke, 2010) that my students created as a class: they wanted to write their own story about the trashed and abandoned lot next to the classroom, about their soon-to-be-garden, with this same construct.

By reading the novel, concepts of empowerment through community were explored. Noticing first the theme that each character in the book had their own specific vegetable, my third graders decided to focus on a particular vegetable themselves. They also recognized other elements and structure of this novel. For example, after character analysis through graphic organizing, students noticed that each character had shared similar personal details: age, religion, ethnic background and their connection to the garden. Modeling Fleischmann's writing structure, my students composed their own personal narratives. Modeling the writing process, editing and revision led to the class publishing a book enthusiastically entitled *Seedkids*.

Earlier, in Chapter 2, Molly tells her story of how the school garden was transformed from a vacant lot to a beautiful garden. Below we present excerpts from two of her classmates, Anandi and Callie, who each write vividly on the transformative power of the abandoned lot into a garden.

> When it was time for morning meeting in *News and Announcements* Ms. Clarke said "Today we are going to plant a garden!"
>
> I raised my hand and asked, "Where?"
>
> Ms. Clarke pointed to the fairly trashed alley way in between the portable and the fence that is in between the alley way and the sidewalk.
>
> "Oh," I said. How could we plant a garden in there, are we going to have to break the concrete or something?
>
> After a long time we went outside into the alley way and started gathering trash. Whew! I thought, we don't have to break through the concrete!
>
> The next day we started planting stuff. Each table grew a different thing, one table grew onions, and another grew broccoli, one grew radishes. We planted the seeds in egg cartons inside until they got big enough to grow outside.
>
> Most of the seeds were small and sphere shaped. Some others were bumpy and looked like a hill on one side and flat on another. They were different colors too like creamy brown, whitish brown, and black.
>
> After a long time the botanist, Madelyn, who we were working with said, "the radishes are ready to be planted outside!" "Yey," I shouted to Molly. "Totally," she shouted back.

When we started collecting dirt inside a wheelbarrow then we dumped it in flower beds. We started digging little rows, but, we could not dig the rows too deep or else the plants might drown. It was fun working with the wet soil, it was not mud but it was still wet, it stuck to our hands.

I turned to Molly and said, "isn't it fun?" "Yea" she laughed back.

Then we went inside and got the radishes out of the egg cartons, it was like holding a newborn baby, we had to be very careful or else they might fall apart. Then we put them in the almost perfect rows. It was raining and I was glad.

(3rd grade student Anandi)

Being an Oregonian I learned how to garden. When I actually started gardening I think I was about 7, which was only 2 years ago.

I've always liked plants, but I wish I could garden more. One day I saw a big pile of dirt at my school. The next day I saw 2 medium sized garden beds. "What is this for?" I wondered. The next day I saw shovels, gloves and watering cans. I knew somebody was going to grow a garden. "But who?" I thought.

I was overjoyed when Ms. Clarke, our 3rd grade teacher said we were going to grow a garden. "But where?" we asked. "Over there" Ms. Clarke answered as she pointed over to the trashed lot next to the classroom. "Oh no!" I thought.

I was very shocked we actually made a garden! We picked up the gross nasty trash. I heard my friend Amahn whisper "eww" as he picked up an old gum wrapper. Then we made the soil. My class and I started with a big pile of dry dry dirt. My friends and I dug up dirt and put it in a wheelbarrow. My friend, Luke and I dumped the dirt in the garden beds. Some of us mixed the dirt with worms, leaves, and hay. We had made compost (good soil). Soon it was time to plant. We dug little holes in rows. Then my class and I carefully put a seed in each hole.

The seeds grew and grew. They were as precious as gold to me. When the first plant came up I almost screamed of joy. At last they were ready to pick. We made salads, teas and more. The plants tasted great! I couldn't believe what we had done. We had made a disgusting trashed lot into a beautiful garden!

(3rd grade student Callie)

Both Anandi and Callie were initially shocked to learn the location of the garden. Along with other students they expressed astonishment that a garden could grow from a trashed lot. These narratives show that children are very aware of their environment. Having participated in the transformation of the schoolyard into a beautiful garden, they intuitively understand their own potential for transformation. Feelings of pride and joy abound in these narratives.

Community Mapping: Getting to Know Our Place

In order to get to know our "place," students made maps of their neighborhoods. At school we made maps of our school grounds, including everything from various buildings, playground structures, to the trees and shrubs. On the map of the school grounds, next to our classroom, between the metal fence boundaries to the sidewalk, many students noticed and included the wooden planter boxes which rested on the cracked concrete. These squares were empty and among bits of trash and debris. What was to become of this space? While reading the book *Seedfolks*, students made the connection that the space and the wooden planter boxes could be potential garden beds for their use. The mapping project combined with the book we were reading, naturally indicated to students that a garden could be planted in this abandoned space on school grounds.

Technical Drawings

In getting ready to be involved with the school garden, my students and I discussed the importance of being scientists in the gardens. We discussed that scientists first observe their surroundings. Then we met the garden coordinator, Madelyn Mickelberry-Morris. I introduced her as our "Resident Botanist." This gave Madelyn a title as an expert. She became a much loved and important character in our garden and classroom work for the year. In the Sabin Garden, Madelyn reminded students to watch with care and concentrate for our first garden experience together. Students were going to create "technical drawings." We spread out, finding individual space to sit and observe. Each student created, with colored pencils, a technical drawing of their chosen observational spot. After each Science Talk and experience with the garden we would write in our Garden Journals. This was a designated space for us to record thoughts and facts concerning, geography, history, writing, reading, math, science, and art. Students also wrote poems inspired by their garden observation, combining their technical observation and use of simile as learned from reading *Seedfolks*. Below, a student sample of poetry is presented.

Sedum

Like beans on a stem
Like weeds flowing in water
Smells like mushrooms

On top of each other
Like a everlasting life that flows
all over the place
starting from one to one
thousand

Student poetry such as this captures the interconnected learning students are capable of, here linking keen observation with poetic sensibility. Figure 11.1 shows a coupling of poem with technical drawing. Beyond the remarkable observational skills and poetic phrasing exhibited in these works is the deeper significance of children getting to know native plants. Most children are capable of identifying over 100 corporate logos but hardly five native plants. Through careful observation, reflection, and personalization through poetry writing, children are creating meaningful relationships with plants and place.

Making Tea with Herbs

At one of the sessions with our Resident Botanist, we harvested herbs such as mint, thyme, lemon balm, and chamomile from the Sabin Garden and made tea. This tea was claimed to be the best tea most students had ever tasted. In fact we

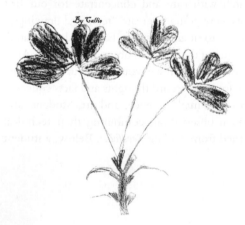

Oxalis
It's like a clover. It's green and hairy. Oxalis is just like a heart.
The stems like a blood vein. It's as salty as seawater
The stem is as rubbery as rain boots. The bottoms
like a palm trees trunk. Guess what it is. It's Oxalis.

By Callie

FIGURE 11.1 Sample Native Plant poetry

made hand-bagged individual tea bags of our "special tea" to give to families as gifts for various holidays and special events throughout the year. Madelyn was a major contributor to the edible garden that we created in the raised beds that were built over the summer. She provided the expert knowledge for planting, sprouting, and caring for seedlings.

Field Guide: Student Ethnobotany Project

In satisfying state standards for history, geography, and science as well as our focus on ethnobotany, we divided the state of Oregon into *eco-regions* and explored the meaning of the word "native" as it applies to both plant and human communities. The eco-regions of Oregon that we studied included: Willamette Valley, Coastal Range, Cascades, Klamath Mountains, Northern Basin, Blue Mountains and Columbia Plateau. This task was jig-sawed so that each student could become an expert of a particular place to share with others in order to put the whole puzzle together regarding the eco-regions of Oregon. Oregon has some of the most diverse eco-regions of any one state, providing rich opportunities for students to explore. Criteria were set so that each team had responsibilities the others needed from them. For example, each team needed to include a map of where their region was in Oregon and discuss the climate of their region. This assignment fueled many activities and products including salt elevation maps but, in particular, a PowerPoint to share their research with one another. The decision for a PowerPoint as means for sharing their found information about Oregon eco-regions with others was one of real world practice. We also had a celebration of our learning. Students created displays about Oregon history, science, and geography. We sold our books, cards, and poetry, and gave garden tours, donating the money. We also invited other classes and community members to participate in our celebration.

A field guide was also developed by the students as part of the eco-region study. A variety of climates in Oregon were investigated, particularly in relationship to plants. This led to an understanding of native plants and how they adapt to the region. Students did technical drawings of native plants. This also led to further Science Talks about why certain plants grew in certain places. Building from their engagement with local native plants in the schoolyard habitat, students broadened their knowledge of what it means for a plant to be "native to a place" through studying diverse eco-regions of Oregon, tying together the study of place, common and Latin names, technical drawing, classification, scale, and cultural uses. The student field guides became beautiful artifacts worthy of sharing.

Garden Journals

Students kept garden journals, recording their ongoing inquiries. These journals gave students space for reflections, poetry, drawings and feelings that came up as they worked. The Sabin Community Native Garden inspired a deeper view of native plants. Stewardship was born for this local place—the Sabin school grounds. During family garden workdays and in-school participation, a fence was built and pathways were improved. Students and parents worked together to assist in protecting this newly cared-for space. In "Seventeen Rules for Sustainable Community," Wendell Berry (1992) urges preservation of ecological integrity by asking how the local needs might be supplied from local sources, including the mutual help of neighbors. This concept was exemplified in the collaboration of students, teachers, families, and the larger community in the Sabin Native Garden.

Bringing it All Together: Nature as Teacher

The technical drawings, Science Talks, and garden time all inspired a connection to our reading and writing concerning *Seedkids*. We tried our hand at using similes to describe the native plants. After keen observation with loupes (magnifying glasses) examining the plant in order to view things unseen at first glance, students began using similes to describe what they saw. One student exclaimed, "Wow, Oregon Grape leaves are as sharp as dinosaur teeth." Another observed, "The sage leaf is like a towel brushed the wrong way."

With *nature* as teacher, my class became enthralled with relevant local issues as we discovered a sense of place. Within the experience of schoolyard map-making and neighborhood map-making, children began to notice the wild in-between spaces and included them on their maps. They marked fences and trees where they had relevant memories with friends or with nature herself. These children were developing a sense of place because they identified more closely with the space that surrounded them. Instead of seeing the natural world as a "backdrop," they were able to recall moments of connection and bring nature into the foreground. The narrow sense of self became a place-based sense of self.

A new way of looking at the world was achieved. We did not only analyze and recognize the beauty of our surroundings, but we created an affinity which inspired awareness that we are part of something grander than ourselves. Using nature as our teacher allowed us to be *in* the conversation versus having a conversation *about* our world. This bond with our natural world is a different way of knowing than rational/scientific study. A sense of nurturing, exploring, and empathy can be developed because of recognition of "place." Our cultivating feelings of connectedness with the schoolyard and school grounds and gardens also helped us build the emotional foundation necessary to understand more abstract concepts of everything being connected to everything else.

Learning Sustainability in the Garden

Incorporating multiple concepts within a simple framework was a high priority in the creation of my integrated *Ethnobotany* curriculum. "Our children need to be fed knowledge and food in more than fragmented parts and pieces. They need to understand the whole process and the interconnection of all things" argues Michael Ableman (2005, p. 179). Capitalizing on the notion that children have a natural affinity for the living world, I structured my curriculum to include school gardens. My curriculum was designed to join ethnic knowledge and plant knowledge accessing various content subject areas. Social studies, history, sciences, reading, writing, and math can be integrated by the uses of a school garden. School gardens are a powerful means of instilling lifelong environmental and nutritional literacy among children as well. Gardens can serve as living laboratories for hands-on exploration and learning, and planting can be a strong component for teaching ecoliteracy. In my experience, classrooms that utilize the outdoors can foster critical thinking skills.

School gardens can be used by teachers to serve a variety of goals. For some, a garden could be an enchantment with nature, feeling the magic of nature. For others, there might be a need to use rationalism when making use of the outdoor classroom. Place-based education embodies systems thinking. Students establish interconnected ideas that allow them to explore the environment throughout the year without barriers to productive, integrated thinking. In my class, the acquisition of knowledge was facilitated through actual doing and interacting, which contributed to the wider world to which the students belonged. My students enjoyed being in the garden and working outdoors in nature. The garden work was not only a medium for them to learn and experiment as scientists, but also their work was seen as beautifying the school and helping the local community. It pleased them to be doing something that was clearly useful and beneficial to others. The aim of place-based education is to ground learning in both local phenomena and in students' lived experiences.

In my classroom, students' personal questions and concerns played a central role in determining what was studied. An ownership emerged. For example, when asked, what was one of the most memorable events garnered from class, one student responded, "Probably seeing our garden in full bloom and knowing our class had changed the world just a little bit by doing what we did." Since students had a chance to actually participate in the creation of their own learning, I saw myself as a co-learner among these budding minds. My responsibility did not lie in my stored knowledge, but rather in my skills and gravitation to acquire outside community experiences and to foster student inquiry about place, the school grounds, the gardens, and the community. Entering the community, students had the chance to participate in service. These interactions were not merely significant academically, but were valuable to their community. Each situation validated students' own existence. Children are primed to learn things that bring them genuine praise and to relish the satisfaction that comes from helping others. Establishing bonds helped create a caring attitude toward nature and each other. In this atmosphere true notions of sustainability began to blossom because they had been watered with integrated curriculum and fed systems thinking as the fertilizer.

Sustainability education ties curricula, subject areas, and processes together rather than necessitate the development of new curricula. Sustainability education helps students learn how to think in whole systems and how to find connections. It teaches students how and when to ask the "big questions," and how to separate the trivial from the important.

Students who use nature as a teacher celebrate wholeness and appreciate the fact that life has more to reveal than human ingenuity has yet discovered. Nature teaches her children to pay attention to the world around them, to respect what they cannot control and to embrace the creativity with which life sustains itself.

What is good for the future of the environment and for communities can be learned in schools. Students who learn nature's principles in gardens and serve their communities through civic participation become more engaged in their studies and score well in diverse subjects, including science, reading and writing, and independent thinking.

Leaving a Legacy: Growing a Legacy Garden

Garden work was a powerful and meaningful task my students engaged with during the 2008 school year. When I asked them what was their most memorable event from their school year, many students responded: "garden work" and "creating the herb spiral." The herb spiral was a garden element implemented based on permaculture concepts. The herb spiral was planted in the spring of 2009 and was dedicated by the class to future students as a "Legacy Garden" (Figure 11.2).

The intention was for students to pass along the knowledge that they can have positive influence on others' holistic experience when they create gardens. My students learned from direct experiences earlier in the year about the benefits of herbs in the preparation of food and medicine, that some plants attract pollinators and that the plants provide a source for ritual for those who use the garden space. "Gardens provide great metaphors for life, the circle of birth and death made palpable because it is seen firsthand, year after year. Working with the soil offers a sense of accomplishment and personal power" (Ableman, 2005, p. 181). Students understood these concepts because of the integrated study of history and science earlier in the year. Students had the experience of harvesting herbs, eating them fresh, and enjoying them together in tea. Planting more herbs by building a herb spiral enabled more students to have these experiences—a truly spiraling effect.

My students also learned about and reflected on the spiral as an effective design strategy for gardening. The spiral design allows an effective use of space as more plants can fit into a smaller area. Plants in the center of the spiral are much easier to reach, and the mounding of the soil means less stooping to harvest. The slopes of the herb spiral face all directions, which create different microclimates of sun

FIGURE 11.2 Legacy Garden constructed by students

exposure and water distribution. Spirals are present everywhere in nature, much more so than the standard rectangle of raised beds. My students learned about the concept of designing gardens in a way that more closely resembled nature's designs.

When it came time to plant, my students did research on the different herbs, noting interesting facts about them and carefully considered each herb's placement in the spiral. To make the herb spiral an educational experience for visitors, my students made identifiable signage for each herb. Their making a technical drawing of the herb helped sharpen their scientific observational skills. This was particularly enjoyable because of their earlier work in creating a Native Plant Field Guide while they studied Oregon's diverse eco-regions. These students reflected deeply upon the legacy they were leaving behind with this positive project. When asked about their thoughts on the importance of the legacy in caring for the Earth, their responses were poetic and insightful:

> Legacy is something to consider when thinking about the Earth, because when you remember something that was good for the Earth and lasted a long time a new generation can use it.

> *(3rd grade student)*

When you have a good idea you should make it a legacy, so people could use your good idea and the Earth will be a better place to live in. Other people could pass on the good idea and then the plants could live better.

(3rd grade student)

The goals of sustainability education are exemplified by the implementation of school gardens as part of the curriculum. Gardens are truly places for communities to come together and also for students to get direct learning experiences. They provide children with an accessible place both to develop long-lasting relationships with nature and, most importantly, to learn about the complex interconnected web of life.

Lewis Elementary School (K–5), Portland, Oregon

Lewis Elementary School in Portland takes pride in the Lewis Outdoor Education Center and in the variety of gardens that have been built on site. The neighborhood community and school parents along with the children and their teachers have ownership of the school and its gardens, where Tim Lauer has served as principal since 2003. As the leader of the school, Lauer has championed the revival of the gardens and the garden program at Lewis.

Lewis is fortunate to have a 40-year history with environmental education. Being within walking distance to Johnson Creek has helped with connecting students to the outdoors. Beginning in 1968, the creek served as a "learning lab" with one of the first environmental education program in Portland Public Schools (PPS). The exposure of students to the outdoors encouraged teachers and the community to talk about better use of the plot of ground west of the school, which, in 1975, resulted in building a fence and enclosing the yard along with the installation of security lights. This area, almost four decades later, still serves as an enclosed garden area. In 1977, the entire staff of Lewis was involved in an in-service class, *School Ground Science,* resulting in teachers coming together with the community to build the first greenhouse, which is still on the grounds and in the last few years has been used as a tool shed. Some of the apple trees that were planted still live in the area designated as the "orchard" on school grounds. Trees and shrubs were planted with support from the district's Green Thumb high school students, and Lewis classes started to grow vegetables and flowers. To support outdoor learning, an Environmental Science aide was hired.

Between 1980 and 1995, the outdoor learning center was used by classes in small groups. Several volunteers, including parents from the community who were often garden specialists, helped continue the uniqueness of the school program. With a National Wildlife Federation grant of $2,500 in 1996, a small pond was put in and more trees were planted. Over the next few years, with teacher retirements and turnover, there was less integration of the gardens in the school as a system. However, volunteers and a 4-H program continued to support an after-school garden club and the outdoor center program was staffed with teachers, parent volunteers, and community members between 1998 and 2002. In 2003, Tim Lauer was hired as principal of the school from within the Portland School District.

In 2004, Professor Pramod Parajuli of Portland State University (PSU), secured a large Environmental Protection Agency grant to work in partnership with schools for a project called FEED (Food-based Environmental Education Design). Parajuli sent a Request for Proposals to all PPS schools in the district, seeking schools interested in establishing or strengthening garden programs. Lewis, a Title 1 school, with over 40% of the 400 students on free or reduced lunch program, was one of the schools that applied for the grant and was selected for partnership. This enabled principal Tim Lauer and his team of teachers and parents to revitalize the garden program. For eight years now, the principal has been committed to the learning gardens and has looked for creative ways to fund a position of coordinator from the school budget. In 2011, the school garden programs are strongly anchored in the school. We requested Principal Lauer to share the story of the school and his involvement as a leader facilitating the flourishing of the learning gardens at Lewis, to cover the following broad topics: designing the landscape; honoring history and tradition; connecting gardens with lunchroom and other green school initiatives. This involved making gardens "visible" in the school; building partnership support for the gardens with higher education, neighbors, and community; facilitating infrastructure support for teachers; and linking gardens with academic content. Below, we capture his story in his own voice.

Partnerships are Key to Success—Tim Lauer, Principal

When I was interviewed for the position of principal in 2003, I was asked by the site council and the Parent Teacher Association for my views about gardening and the outdoor center that had been revived by the parents at the time. The community felt this was an important part of Lewis, its history, and its ethic of stewardship. Since I joined Lewis eight years ago, we have had a number of very active parents who have made a commitment and volunteer innumerable hours helping with writing grants, building raised beds, and supporting the garden coordinator. We have work parties, community-building activities, and regularly bring students to the gardens and our outdoor learning center.

Building Partnerships

In 2004, we learned about Portland State University's (PSU) FEED program through a request for proposal that was sent to the district by Dr. Pramod Parajuli. One of our parents, Amy Spring, who also has an appointment at the university, is a grant writer and in collaboration with the teachers and the site council, we wrote a proposal. This was a "natural" for us since we wanted to revive our school gardens. We were one of three schools in Portland Public School District that was awarded the FEED grant. This three-year grant provided graduate assistant support, visits to the Bay area to see school gardens, staff development in garden-based education, and supplies to build gardens. One of the requirements of the grant was that there had to be a school garden committee that at least included two parents, two teachers, and the principal. Besides Amy Spring, we had another parent, Gregory McNaughton, serve on the committee along with teachers Sarah Jones and Kathy Gould who were excited to be part of the team but they were cautious at first. Gregory McNaughton, who was himself a gardener, was particularly interested and became the primary contact for the work that we began in partnership with PSU.

This partnership with PSU was key to maintaining the interest in the school garden program and it provided the foundation and momentum for our work: PSU's support, in terms of graduate assistants from the *Ecology, Culture, and Learning* program—Catie Pazandak and then Angela LeVan—was invaluable. We also had a large number of PSU students involved in service-learning at our school. This tie with higher education was important for sustained support.

Supporting Teachers

As we got the grant, I invested in an AmeriCorp position for a garden coordinator for about 30 hours a week. The FEED grant supported the committee to undertake a trip with the other two FEED schools (Buckman and Edwards Elementary) to the Bay area with PSU faculty. The four-day fully funded trip took our team to visit the Martin Luther King Middle School in Berkeley where Alice Waters had the *Edible Schoolyard* program. Teachers came back excited because they had observed students cooking in the cafeteria and serving food. There were tables covered with colorful tablecloths and flowers as center-pieces. There were no Styrofoam throw-aways. It was also exciting for them to visit the *LifeLab* program at the University of California at Santa Cruz, which has a strong curriculum component that teachers particularly appreciated.

The team also visited several schools in the San Francisco Unified School District: Alice Fong Yu (K–8), Tulle Elk (K–5), and Ulloa (K–5). Meeting teachers, principals, and parents in these schools and listening to how these gardens supported the learning environment especially encouraged the teachers on our team to have confidence that gardens can be integrated in the school. They saw children actually having curricular sessions in the school gardens. I believe that these sorts of site visits are crucial for educators—they make the garden program more tangible. When teachers hear from other educators, it gives them confidence to undertake something new. Having undertaken the trip also helped the team bond with other FEED teams, and helped jump-start the program and get it going. Having parents and teachers join together and do a field trip was a good strategy since it made us feel that we were not alone.

At Lewis, the teachers self-selected to participate in the FEED program. There was general interest among our teachers to design a couple of lesson plans and tie in the lessons with the science curriculum as they wanted to take students out in the gardens. Staff development sessions with the other FEED school teachers was critical in exchanging ideas and having us all feel that we were part of something "bigger" in garden-based learning. We were not isolated.

When we started, we also did soil testing; the grant paid for that. We felt that doing this was important if we were going to grow food. I wanted to be able to defend what we were doing.

The original core group stayed with the program for the three years of the grant as other teachers got interested in trying new activities in the garden. I also looked for funding in my own budget for an AmeriCorp position as coordinator of the garden so that we ourselves could make a commitment and have a staff person from the school devoted to this work.

I strongly believe that a garden coordinator is essential for learning gardens. It is difficult for a classroom teacher with all the things that they have to do to safely take students into the garden. It is better to take them outdoors in small groups. We do not want 20 kids handling shovels all at once. The AmeriCorp coordinator spent almost 30 hours a week on the garden project. The support from PSU in the form of a graduate assistant significantly helped with working on curriculum and lesson plans and taking the students outdoors.

To understand the scope of the work and how important partnerships are, let me share what Angela LeVan did over a period of two years, serving as a graduate assistant from PSU and putting about 20 hours per week into the garden-based program; we had an AmeriCorp coordinator working with her as well. Angela supported 150 students in grades 3, 4 and 5, integrating garden-based and place-based curriculum. This amounts to almost 300 hours of instructional support. Since she had a teaching license, she had credibility with the teachers. During that time, Lewis also built several raised beds and revived others with work parties involving parents and community. An after-school Garden Club was started; this attracted about 15 students each year. Between 2004 and 2007, we planted and grew food on-site using 12 garden plots. Students went on field trips to Jean's Farm and the Learning Gardens Laboratory, each within walking distance of the school. This gave significant exposure to the students about why learning how to grow food matters.

Making Gardens Visible

Besides our regular work parties, some of the other very exciting activities that we have had each year are the Earth Day celebration and the Martin Luther King service day celebration. Volunteers including parents and grandparents are ready to work shoulder to shoulder to improve our school grounds, help with sheet

FIGURE 11.3 Lewis cob bench with sustainability features

mulching, plantings, building beds, harvesting, selling seedlings that we grow in our greenhouse, and cooking and eating food together.

In 2006, we planned a cob bench (Figure 11.3) and the following year we started building an outdoor classroom.

Cob is the combination of sand, straw, and clay. Children mix it with their feet and apply it like plaster. Urbanite (broken up pieces of concrete) was used for the foundation. The oval shape of the cob bench provides a natural gathering space for students entering the garden from the school building. The rounded edges are soft and graceful. The outdoor classroom uses a typical truss construction with a green roof to absorb rainfall. In a maritime climate there can be significant winter rain. Water-loving plants grow on the roof to absorb some of this moisture and decrease run-off into storm water drains, which protects the river. Students can watch as extra water from the eco-roof drips down a suspended chain, into a rain barrel, and finally seeps into an adjacent rain garden. Every drop of water is used to support green plant life and create learning opportunities.

Without the garden coordinators Lewis could not have done this since they also helped organize a fund-raiser dance and concert and worked with Village Building Convergence (VBC) to get support. Matt Bibbeau, another PSU graduate student, was instrumental, along with Angela LeVan, in pulling together the community around a common interest for an outdoor classroom that we built with VBC support. This was especially critical to our success since he was able to do much of the paper work in getting permits from the City for building the outdoor classroom. He helped us get two grants in two years through City Repair's annual VBC program. The VBC is an annual "place-making" festival that brings together the passion and interest of the Portland community to be involved in hands-on projects that support sustainable living. Schools can ask for this support by sending a proposal.

Having the outdoor classroom has been a great boon to our school. Parents who were carpenters and builders and landscape designers worked with the staff of VBC and, to this day, students who were involved in building the cob bench and the outdoor classroom come back to us from their middle schools from time to time because they say they have wonderful memories of being involved in the outdoor program that they helped construct. These structures also include an eco-roof and barrels for water harvesting. These spaces are used for academic learning and also for just sitting calmly and enjoying the outdoors.

Between 2005 and 2008, each year I have been able to fund one AmeriCorp position—the third one was Julia Hamlin whom we hired in 2008 and she has now been at the school for three years. It has been good to have continuity with her—for the past two years I have supported her through the school budget and also partly through PTA fund-raising. We were able to create a Community Volunteer Coordinator position that has broadened the scope of her work which includes being a garden coordinator.

Designing the Landscape

If other principals are thinking of starting and/or supporting gardens, I would say, first look at the grounds, decide where the community wants to go. What do you want to transform? Do a visioning process with your community. We did that with our parents and invited neighbors. We also took advantage of a University of Oregon graduate student who took up this project as her capstone project and supported the school in developing a master plan. We discussed the kinds of beds that were already there, and what we liked or wanted to change about them. We wanted our school to be a good neighbor. When we were going to plant trees in our parking strip, we involved the neighbors who had concerns about not reducing the school's parking lot. So, by inviting the neighborhood association, we ensure that they have correct information and then they have ownership and support what we do. This also applies to the different kinds of garden spaces, ensuring that the gardens are well maintained so that they do not become an eye-sore, especially when school is out for summer. For instance, we had a north-east area with some old hedges that were planted decades ago and did not look welcoming. So we dug up the entire area and cultivated the grounds to create beautiful learning gardens. The neighbors are happy with us. Another thing to keep in mind is the positioning of the garden for easy access from the classroom and, if possible, from the cafeteria. In our case, at Lewis, teachers have good and easy access to the garden spaces and the cob bench.

Visibility

We have also been good at telling our story. We invite the local newspaper, and the neighborhood newspaper, and let the reporters talk to teachers, parents, and the children. Not only do these things tell our story to the community, but we also document what we do. I place photos and images on the school's website promptly after events. We have a garden section on the website just to let people know what is going on with our garden program. In many ways this also keeps our history alive. Below is what we state on our website:

> The Lewis Outdoor Education Center provides students the opportunity to learn about water conservation, gardening, native plants, and composting. The Outdoor Center includes an outdoor classroom, a greenhouse, four raised beds, rain barrels, a cob bench and apple trees. The Lewis Learning Gardens curriculum encompasses planting, growing, harvesting, and preparing of foods. Grades K–5 learn about various topics such as seed saving, herbs, organic gardening, and growing starts.
>
> Lewis Gardening Summer Camp focuses on harvesting, preparing, and eating foods from the garden as well as talking about the nutritional value of plants in our diet. There are numerous community partnerships that make the environmental education program possible. Our Garden Coordinator establishes and teaches the environmental programs for the Outdoor Center and the Learning Gardens. The Garden Coordinator also manages new garden projects, teaches the summer gardening camp, and coordinates volunteer efforts such as Earth Day, the 3rd grade Going Green Team for composting lunches, and the Tray Washing program. Lewis also involves Portland's Metro educators to present puppet shows to K–2 and talks to 3–5 focused on recycling and the health of our planet.

There is an important connection of the gardens to the general ambience of the school. The main entrance of the school must be welcoming both inside and outside the school. In front of the building, we have flower gardens, native gardens and a butterfly garden which are invitational for those who approach the school. Close to the lunch room, in an enclosed, fenced area, we have several food gardens. Our fields barely had trees, so a few years back we planted trees to provide shade for families in the summer. Our grounds are attractive play areas. We also have bicycle racks and in our parking strip we planted nine trees in 2010 that will not only help with water runoff but also provide shade. We always involve our neighbors in decisions and keep them informed.

We have also organized several community events. A garden fair brings the community to school and serves as a fundraiser. For Martin Luther King's birthday, we have a day of service when parents bring their children to the school and we often have grounds clean up and a number of community-building activities where we cook simple food, and serve food from our gardens and local farms. Some of the grandparents accompany their grandchildren and tell them stories of when they went to Lewis and what the school was like. For me it is a great joy to see intergenerational communities come together. These community-building activities are important to help the public feel their schools belong to them. The gardens are an attraction as they also provide aesthetic value to the school.

Connecting Gardens with Lunchroom

Our learning gardens link with our other green initiatives such as cafeteria composting. We have been doing lunch-time composting for scraps from the salad bar. Kids take turns collecting the compost and our garden and community outreach coordinator leads the effort. The third grade students run this program. They are called the "green team." Food scrap composting at first was a challenge in terms of getting the system set up to work smoothly. We do it in a responsible way so that it does not lead to rodent problems. When the district sees that it is consistently handled well, and done well, then they allow us to continue to compost. And, there have been no complaints from neighbors, either.

We do not use Styrofoam plates any more. Our parents decided that they would get washable lunch trays and silverware. So now they, along with students, take turns to wash the trays. When we are not throwing away 175–200 styrofoam plates daily, we can feel the impact of our actions. And, frankly, the district office is a partner and supports these efforts toward greening our schools. When we use trays and wash our milk cartons, the district saves $3,400 each year, and we help reduce waste. Instead of the two dumpsters we used to have, we now have only one dumpster. That is a saving for the school district. We are certified as an Oregon Green Premier School. Collectively, because garden programs are linked to all these efforts, there is an attitude of pride in our stewardship at Lewis. Also, there is a commitment and an attitude of ownership that we belong to this place, this community.

The students who graduate have wonderful memories of music, drama, production, art and gardens, including the building of the cob bench. Since 2006, the cob bench project has been very special. The children who helped build it love it because it was very organic and they have memories of constructing it, utilizing it, and sitting on it writing their stories, or quietly reading books.

They remember the wind chimes or the bird-bath that they made as part of their art activities. They remember their salad beds and the herbs they planted. They show up to just hang out with their past teachers. At the end of the day, that's what matters, that's what makes a difference. They have good memories of being here. When this happens, then you are doing something right.

Linking Gardens with Academics: Gardening and Environmental Education Curriculum, 2010–2011

We integrate the garden with the regular curriculum and utilize a block system, with a total of six blocks. Each block is five weeks long. Each class has gardening/environmental education class once a week for five weeks, during two different blocks over the course of the year. Every student (400 total) has the opportunity to visit the garden a total of ten times during the year. The garden coordinator works with half the class, each for 30–45 minutes. During this time the classroom teacher has the other half of the class and can work on writing/reading or any other subject that also benefits from small class sizes. More vocabulary words are integrated, which connect to the content they are learning in their classrooms. The curricular themes, by grade level, are as follows:

Kindergarten

- Trees/Leaves
- Parts of a tree
- Seasons
- Worms

First Grade

- Insects/Pollination
- Seeds
- Parts of a flower

Second Grade

- Animals/Habitat
- Weather
- Resources/Recycling

Third Grade

- Compost—Third Grade Compost Team
- Organic Gardening/Soil
- Plants/Plant Life Cycle/Sugar Snap Peas
- Plant seeds/Measure growth

- Chefs Move to School (partner chef David Anderson) cooking series
- PPS Local Lunch/Harvest of the Month: food origin/local food

Fourth Grade

- Garden Planning/Greenhouse Operations: Three Sisters Garden
- Native Plants: Lewis and Clark/Native Americans
- Resource Conservation/Earth Footprint

Fifth Grade

- Food Web/Food Chain
- Water Cycle

The most important thing about all of this is that it is about learning. Our students have creative writing science journals. Julia Hamlin meets with all the teachers at the beginning of the year and together with the teachers plans the entire year's curriculum based on what teachers are covering in class and how seasonally appropriate tie-ins can be made. She takes students out into the garden in small groups. On a nice warm day, they can use the cob bench; many teachers, themselves, use the space for creative writing.

I have evidence about children's knowledge and understanding of steward-ship; it is incredible. Their wisdom and knowledge does not come out of the blue. While they are particularly learning science, they are also developing an ethic, an understanding, of sustainability. Being good stewards and taking care of things that we cherish, we value. And, for many of our families, these values are reinforced at home.

Children and teachers respond to the sounds of birds that the gardens and fruit trees bring. Several classes are able to look out of the windows at the flowers and plants that are growing in the gardens—it is quite calming for the students and I have noticed their behavior change when they sit quietly in the gardens or work with one another planting and growing.

It helps most that our teachers have ownership, they believe in the school gardens and the other efforts we are making in greening our school. And, parents are happy. Our school is known for its garden and arts program for which several families want to transfer to our school from other neighborhoods. Parents and the community explicitly value our educational philosophy and approach.

Lewis has a holistic approach to education. Art, music, nature, school grounds all tie together. I don't want the school to be just the bricks and the walls: the school is an "idea." The school is a place where you can have lots of different interests and learn by engaging with these interests. Even though Lewis is a Title 1 school with a fairly large number of students on free and reduced lunch, our wealth lies in the kind of holistic education we offer. Among other activities, our school gardens are significantly tied to the identity of Lewis.

San Francisco Unified School District, San Francisco, California

In November 2003 and again in November 2006, San Francisco voters passed two local school bonds for San Francisco Unified School Districts (SFUSD). Intended to bring the district in compliance with the Americans with Disabilities Act, the SFUSD facilities bonds have also been issued for the overall improvement of school facilities and sites.

The 2003 bond authorized the issuance of $295 million in local bond funds and the 2006 bond authorized the issuance of $450 million in local bond funds; together, these measures, when augmented by state matching funds, were proposed to fund a total of 96 facilities on 91 school sites. The San Francisco Green Schools Alliance (SFGSA) is an advocacy group that provides professional development for teachers, and horticultural supplies. The SFGSA was successful in advocating for about $7 million for greening of the schoolyards at the proposed sites. As SFGSA reports, 45 elementary schools were slated to receive $150,000 each to green their schoolyards after a rigorous planning and construction process. Twelve middle and high schools also received funding through a grant application process for their gardens and schoolyards. Teachers, principals, students, parents, and community members got together to make proposals for the design of the school grounds. When there is ownership, the community and all the stake-holders buy into the process and the outcome, thus ensuring long-term sustainability.

Working closely with the SFUSD Bond Greening Program that manages the bond projects, the SFGSA supports the programmatic goals of planning, designing, and constructing the outdoor spaces. While the funds cannot be used for maintenance purposes, the SFGSA has been helping the process of creating a vision at each site for a garden program and also ensuring that teachers are in a position to integrate the gardens into their teaching.

This effort needs to be seen in the broader context of sustainability that SFUSD is supporting. There is a director who is managing and directing the Campaign to Green our Schools. Thus, the gardening programs have a broad support base and interest that includes improving energy efficiency, planting of trees, addressing rainwater overflow issues and harvesting rainwater, all of which are directly or indirectly linked with school gardening efforts.

The SFGSA has supported the establishment of gardens at four schools that specifically deal with rainwater harvesting in their gardens. A watershed stewardship curriculum is also designed and teachers are trained on how to use this curriculum. With Friends of the Urban Forest, the SFGSA has created an initiative: 2012 by 2012—that is, planting 2012 trees on school grounds by the year 2012.

In 2005, the Commission on the Environment in San Francisco advocated that the Mayor's office work with SFGSA to establish a garden at every school.

The Commission adopted a resolution highlighting a number of compounding benefits of school learning gardens, ranging from improved student engagement and academic achievement to increased student sociability and from decreased toxicity in the environment to increased property values and quality of life.

We requested Carlos Garcia, superintendent of San Francisco Unified School District to share his thoughts on what has prompted SFUSD to take on so many gardens at school sites; how the gardens are provided infrastructure support such as curricula, design, and maintenance; and what his views are as a leader of a large organization managing almost 100 school sites. Below, he shares his perspectives.

Infrastructure Support for School Gardens—Carlos Garcia, Superintendent

I was a principal in the SFUSD in the late 1980s to early 1990s, so I was familiar with the workings of the district and had a deep commitment to the children since a large percentage were non white students, especially Hispanic and Asian. When I was offered the position of superintendent in Las Vegas, I left and then moved on from there to McGraw Hill where I worked for a few years. I became superintendent of SFUSD four years ago, in 2007.

The school district passed two bonds over the years that included the creation of gardens. There was a lot of interest in food in the city and it is critical that urban children realize where their food comes from. By growing food they value what is entailed in growing food.

Our curriculum is integrated with gardens: writing, science, health, math are all tied to gardens. There is a state-wide California curriculum that ties the standards to gardens so it is easy for teachers who are interested to access the curriculum for integration.

I believe that gardens can serve as a springboard for learning that is relevant to students. That's the spirit. It isn't simply about a garden being a "nice" place to be. No, it is more of a lifelab. Students learn about their own life through the use of gardens. It is about joyful learning. Learning is hands-on, students can see the end product which could be a crop or simply a flower. For me, a garden is an illustration of what education ought to be like. It is place-based education, a good way for students to make education relevant.

I am a big supporter of school gardens. To know why, you have to understand where I come from, what my culture and my upbringing are. Like many of our students of today who are children of migrant workers, I too am a child of migrants.

My father migrated to the United States from Mexico through the Bracero program during the war. For immigrants from Mexico, this program sponsored by the USA permitted Mexican farmers to do agriculture here. While my father was a migrant worker, moving from place to place, I was born in Chicago.

But then, my father took the family back to Mexico when I was only two months old. I returned to California when I was five years old. And we moved a lot. In fact, I picked lemons in Goleto near Santa Barbara in the summer. Only migrants were in the fields. The conditions of the migrant workers were awful as has been documented. I witnessed poor mothers working in the fields with children. They were given no dignity. Once you work the field you realize there was no place for people to go to the bathroom. No facilities. I met César Chavéz when I was in college. When the farm labor movement started, I joined. The supporters of César Chavéz won in the sense that at least bathrooms for workers were installed and are now required.

We valued education. After I graduated and taught, I became a principal. I moved to a school district in Watsonville near Monterey Bay, where I was an elementary principal in 1983. There, I had teachers who came to me with the idea of a garden. The question they asked was: What if we create a lifelab outdoors? Since most parents are migrant workers, their children who are students in our school can be validated. The teachers said that it would give them an opportunity to learn from the parents and that they would integrate gardens in the science, writing, and mathematics curriculum. This was their way to honor the field workers in the schools, since large numbers of our parents were migrant workers.

An impetus for this initiative also came from the fact that some teachers were working with the *LifeLab* program at the University of California at Santa Cruz. Some had student teachers there. There was one teacher, Terrie Marchise, who was a champion of the idea and wanted to teach science to all grades instead of being a fifth grade teacher in a self-contained classroom. I liked the idea but there were many teachers who resisted the idea. However, they were willing to give it a try. This actually worked really well, and it was my first major experience as a leader, to support teachers' passions as long as they were linking the garden to the curriculum.

Then, I left for a position as middle school principal in San Francisco Unified School District, and then to district level and superintendent positions in Fresno, Sanger, and Clark County, Las Vegas, after which, I returned to San Francisco as superintendent. Gardens were in schools in Fresno already. I saw my job as superintendent to allow their initiatives to move forward and to get out of the way.

At SFUSD, it has been a real joy for me to see how schools are working in partnership with various organizations to address food and health issues through gardens. There are several partners and non-profit organizations that provide support to schools. They have had a long-term affiliation with us. At Ida Wells and at Martin Luther King Jr. middle schools and many others, Urban Sprouts works in close relationship with teachers, especially science teachers. They work with our diverse and multicultural populations. Recently, we have a Farmer-in-Residence program that encourages diverse families to be connected to our schools. The gardens are also linked with the other initiatives that we have related to "farm-to-school" that supports healthy food for children. In addition, we have the Harvest of the Month where we work with local farmers to bring fresh locally grown food into our cafeteria. This is popular across all our schools.

Another important way that we provide support to the gardens is having a visible office at the district level that specifically deals with sustainability issues. There is a Director of Sustainability supported by the city and the school district. Such visibility makes a statement that sustainability is important in our district. In 2005, the Commission for the Environment passed a resolution advocating for a garden in every school. The resolution document is publicly available.

I believe that we can use gardens as an educational tool, make learning joyful, fun, and relevant. Every child eats food. Gardens create connections between what students see in the market and what they learn in science or writing. The educational value is most important—gardens make education relevant. Students get to do all sorts of experiments in life sciences by actually using plants. We all know, as educators, that students learn more through experiential learning—so gardens ensure that there is another instructional tool. This point is really crucial: it is one thing to have a garden and another to use it as an instructional tool. Gardens are more like lifelabs. Just as we have science labs, gardens are instrumental in teaching curriculum and motivating kids to learn. I feel that teachers like using the gardens as long as they can be shown that this is an easy way to integrate science in the curriculum. Let us give the kids back their curiosity. Let us get kids motivated.

While it is true that some schools might not be able to afford a garden, I would say, do it as a community outreach with community-based organizations and local foundations. Cities and counties might want to network with parks and recreation, and use them to co-sponsor gardens. We now have a rooftop garden at Sanchez Elementary School. This is the work of partnerships. If schools work with communities, then the community feels invested and also special. There can be a domino effect with community members supporting schools and then it affects achievement. For parents who work in the fields, gardens bring them into the community, and validate them for who they are.

I would not want people to think of gardens as "just a garden." Rather, I want them to think of gardens as being better than a textbook. Often, the superintendent might be the last to find out that there is a garden at a school in the district and I am fine with that. I do not underestimate that small groups of people can do wonders. Local communities facilitate gardens on site the best since they are the ones who know their schools the best. No superintendent in a school district office can create a master plan for what the gardens in all their schools should be like. We can provide guidelines but we cannot enforce a homogenous approach to gardens. We were lucky to have passed the bonds, which allow us to individually design the garden programs at school sites.

In closing, to me, school gardens are an academic issue. Students appreciate good food, if they grow it. And the other thing that is an interesting outcome: in our fast-paced world, gardens also teach us a lot of patience. Kids get excited, they work hard, and develop a good work ethic when they are engaged hands-on with life, real life.

Conclusion

The stories shared here capture the robust potential of learning gardens as seen through the lived experiences of a teacher, a principal, and a superintendent. From the integrated curriculum and imaginative projects described by teacher Deziré Clarke, to the community partnerships highlighted by principal Tim Lauer, to the strong district-level institutional support advocated by superintendent Carlos Garcia, these diverse voices emphasize the transformational educational possibilities that emerge from gardens grown from the living soil of the school grounds. From the soil springs forth vibrant life and lifelong learning.

Deziré Clarke's story describes development of learning gardens curricula in a single classroom. This can be inspirational especially for teachers interested in beginning to "green" learning "one classroom at a time." The integrated ethnobotany curriculum designed by teacher Deziré Clarke seamlessly links the study of place to valuing of diversity, and encourages students to perceive the interconnected nature of life in profound ways. Observing native plants in their schoolyard habitats, students connect deeply with place. Studying diverse native plants from more distant Oregon eco-regions, students learn about flora and fauna and contextualize their locale within broader bioregional state geography. The integration of gardening and writing stands out as a particularly creative academic linkage. Gardens are often used for grounding science and math literacy, but as Clarke demonstrates, they can also serve as inspiration for creative writing. Central to Clarke's development of integrated curriculum are the following convictions:

- I have an open-ended approach to teaching but I structure my curriculum in ways that pull all students into the conversation. I wanted to weave gardens into the third grade subjects that I teach; there is no particular book that can teach one how to do that, step-by-step, for everyone, because gardens are so season-specific and also schools are community-specific. But there are many school garden books that are inspiring and encourage us to try as teachers.
- Gardens are truly places for communities to come together and also for students to get direct learning experiences. They provide children with an accessible place both to develop long-lasting relationships with nature and, most importantly, to learn about the complex interconnected web of life.

Clarke's visualization of whole-child learning presents a clear picture of integrated curriculum design, and echoes the ecological childhood development models articulated by Bronfenbrenner (1979) as well as the embedded systems models described by Capra (1996, 2002). The linkage between embedded ecological living systems and interconnected whole-child learning emphasizes a shift from mechanistic models toward ecological models as educational guides. Conceived as more than a collection of assorted subject areas, the school is redefined as an ecological entity filled with interdependencies and interconnections. This vision leads directly to Principal Tim Lauer's exposition of the importance of partnerships in developing school-wide learning gardens.

From Principal Tim Lauer we learn how an entire school can come together to use gardens as adjunct outdoor learning centers. Lauer highlights the need for schools to reach out beyond the boundaries of the school building—to parents, neighbors, and universities. His story illustrates the value of permeable school walls. Not only does active learning occur in the gardens, deep community partnerships are forged as well. Lauer emphasizes the following:

- I don't want the school to be just the bricks and the walls but the school is an "idea." The school is a place where you can have lots of different interests and learn by engaging with these interests. It helps most that our teachers have ownership, they believe in the school gardens and the other efforts we are making in greening our school.
- I have evidence about children's knowledge and understanding of stewardship; it is incredible. Their wisdom and knowledge does not come out of the blue. While they are particularly learning science, they are also developing an ethic, an understanding, of sustainability. Being good stewards and taking care of things is what we cherish, we value. And, for many of our families, these values are reinforced at home.

Of particular significance in Lauer's narrative is the need to provide support to teachers. He elaborates on several support mechanisms, including taking a field trip to visit established school gardens, fundraising to hire a garden coordinator who can bridge between the classroom and the garden, connecting with local universities to develop larger-scale projects, and inviting parents and grandparents into the educational atmosphere. The role of the school in the community and the community in the school shines through the narrative, as in the example of planting a beautiful butterfly garden near the school entrance to create a welcoming atmosphere, or in the many stories of parent, university student, and community volunteers chipping in to make steady improvements to the outdoor learning center. For Lauer, it takes an entire community to grow a garden and to educate a child.

Finally, Superintendent Carlos Garcia shared the cultural significance of gardens as a way for schools to reach out to marginalized communities. Working with students of migrant workers, Garcia emphasizes that the garden is a valuable linkage that validates the agricultural knowledge possessed by migrant workers. In Garcia's vision, schools can use gardens to make learning meaningful for students. Migrants' grounded soil knowledge is neither discarded nor devalued; rather, it is valued in tandem with the study of the 3 Rs. Gardens thus provide a bridge linking diverse ways of knowing. Garcia contends that gardens are first and foremost an innovative educational tool. In his words:

- I would not want people to think of gardens as "just a garden." Rather, I want them to think of gardens as being better than a textbook.
- To me, school gardens are an academic issue.
- I believe that we can use gardens as an educational tool, make learning joyful, fun, and relevant. Every child eats food. Gardens create connections between what students see in the market and what they get in science or writing. The educational value is most important—gardens make education relevant.

Additionally, Garcia emphasizes the value of extra-curricular lessons learned from gardens, including joy, wonder, work ethic, and patience. His passion for helping kids connect with life is apparent. As the leader of a large school district, he supports grass-roots initiation of gardens.

While each practitioner shared concrete steps that can be taken at a number of levels, a theme running throughout all of their remarks relates to an understanding of schools as more than machines. Indeed, each practitioner elaborates a view of schools as interconnected living organisms carefully attuned to place, in which curiosity and wonder abound along with a respect for and valuing of diversity in all its forms. The gardens provide an experiential and sensory component that bolsters academic learning, and impart a seasonal rhythm and living scale to education. Rather than referencing mechanistic metaphors in organizing learning, each of the above voices of educators speaks to a transition toward ecological models as guides for sustainability education where practice and theory are intertwined and mutually supportive of life's learnings in education.

12

MOVING FORWARD

This book has been an invitation to educators to join us in the work of bringing life to schools. Growing gardens is one joyous method that brings delicious dividends. While the examples sprinkled throughout the text speak for themselves and the practitioner narratives in Chapter 11 provide a rich array of opportunities, in this final chapter, we present lessons learned based on our interactions with students, teachers, principals, and others in school districts across the country followed by a reiteration of the role of learning gardens as academic venues. We end the book with a final discussion of living soil as the theoretical foundation of learning gardens and sustainability education guided by an ecological paradigm illustrated by seven principles that link pedagogy with pedology.

Lessons Learned

To begin, we synthesize the accumulated practical wisdom expressed throughout the text via examples and vignettes. There are various entry points for teachers and principals to start and/or support gardens on their school sites as a practical way to bring life to schools. The key lessons are to be flexible. There is no one design or method for ways in which the learning gardens are built or used for learning. The process, like all change processes, is non-linear, iterative, and dynamic. Thus, being in touch with local ecological and cultural phenomena is essential. Every school is unique in terms of its bioregion, history, seasonal and cultural dimensions. Therefore, each school garden will vary in design.

Innovators do well to follow a foundational permaculture principle: observe and interact. By surveying the school grounds for physical phenomena, educators can glean practical information, such as where the sun shines brightest, where the wind blows steadily, where children naturally congregate, and unfortunately where rubbish and debris accumulate. Additionally, through broader observation, educators may garner insights into less tangible aspects of the school grounds such as neighboring community dynamics. From these insights they may determine how a school garden might fit into a larger community context. For example, there may be an adjacent park, or community center, or community garden that may be a valuable resource and partner. In the movement toward sustainability in general, partnerships are critical, as the go-it-alone mentality of individualism is implicated in the un-sustainability crisis. To observe *and* interact requires that we cultivate relationships with place, with community, with soil, and so forth. When we couple observation with interaction, we move from the illusion of objective study toward grounded, subjective, experiential engagement. Observation and interaction are keys to understanding the nuances of one's bioregion and sociocultural context. All three practitioner examples in Chapter 11 reference observation, interaction, and collaborative partnership as keys to the success of learning gardens.

Academic Learning

Gardens support academic learning as a rejuvenating natural context. In Chapter 11, Principal Tim Lauer notes that teachers often enjoy using the outdoor learning space simply as a contemplative and serene site for learning. Similarly, Superintendent Carlos Garcia has observed the calming effects of learning gardens on students. Students and teachers alike are drawn to the garden. Environments that are calming and contemplative encourage focus.

In contrast to typical modern education conducted inside under fluorescent lights and gazing into flickering computer monitors, gardens provide nourishment for mind and body. The seven principles that link pedagogy to pedology provide an outline of how academic learning is integrated through learning gardens. Below we briefly highlight practical examples of academic learning represented by each principle.

In Chapter 4, "Cultivating a Sense of Place," we gave an example of a routine neighborhood or garden walk that students and teachers used as a way to connect with place. Much can be learned from such a "simple" activity when students are guided by a thoughtful and creative teacher. For example, students may learn the names of native plants, observe changes in the seasons, or notice flowers unfurling. Repeated visits encourage perception of relationships over time—a key component of the scientific and historical study. Much is learned, for instance, by returning to observe a tree over time: the tree begins to be seen as more than a collection of bark, leaves, and roots, it becomes a member of an interconnected biotic community. By contrast, looking at a textbook diagram of a tree, one would be mystified as to how it fits into a larger ecological or social context. Why, for example, would a tree create thousands of seeds each year, when only a handful over a lifetime will ever grow into trees? The answer can readily be understood through repeated contact, as one sees squirrels harvesting acorns or walnuts. In the narrative provided in Chapter 11 by teacher Deziré Clarke, we saw how students made detailed technical drawings of native flora found directly on their school grounds. These practices keenly illustrate a connection between academic learning and a connection to place via learning gardens.

In Chapter 5, "Fostering Curiosity and Wonder," we discussed the centrality of an active inquisitive mind to learning. Links between curiosity, wonder, and critical thinking abound in a school garden. Just one step out of the monotony of a modern school into the diverse stimuli of an outdoor learning environment fills one with motivation to investigate the "simple" questions: what is here and why? The contingency of knowledge sparks more questions: why is grass green, and not some other color? There can be no better venue for fostering curiosity and wonder than the multifaceted, ever-changing milieu of learning gardens. Watching as plants emerge from tiny black seeds, for example, is an experience of lasting wonder even to seasoned adult gardeners. As Principal Tim Lauer remarks, when they are engaged in the learning gardens, children's insights are astounding; while they might be specifically learning science, children are simultaneously developing an ethic of care and curiosity, and an interest in preserving the things they come to cherish through contact with nature in the gardens.

In Chapter 6, "Discovering Rhythm and Scale," we described the use of square-foot gardening as a technique for translating two-dimensional geometry into the garden. Following a seasonal calendar encourages students to associate time with what is occurring in their local environment, rather than merely with holidays or weekends. More broadly speaking, discovering rhythm and scale relates to development of a pattern literacy, with which students may begin to understand nature's language, a key component of the emerging field of biomimicry. Building the legacy herb spiral in teacher Deziré Clarke's class, students constructed and planted a three-dimensional spiral garden, learning about the interrelationships between aspect, slope, and soil drainage and specific herb's sunlight, water needs, and growth habits. Connecting the abstract study of three-dimensional geometry to the grounded review of specific plant needs creates substantial academic learning.

In Chapter 7, "Valuing Biocultural Diversity," we highlighted use of a food connections chart to explore various biological, ecological, sociological, and cultural aspects of food. As the students in teacher Deziré Clarke's class found, there can be many names for one plant, an observation that indicates multiple concerns to be carried forward in any name. Planting diverse food crops in learning gardens opens up a conversation on diversity that links biological and cultural diversity and encourages cross-cultural sharing. As Superintendent Carlos Garcia explains in Chapter 11, gardens provide a meaningful cross-cultural connection that welcomes otherwise marginalized communities into the school atmosphere.

In the multicultural gardens that we see emerging within and adjacent to school gardens, a vibrant educational dialogue is taking shape that considers difference as a valuable contribution to creating unity in diversity.

In Chapter 8, "Embracing Practical Experience," we gave the example of building a compost heap as a typical garden activity that lends itself well to experiential learning. But we also highlighted that experiential learning means more than just going outside and doing something. Learning from experience is contingent upon critical reflection after the completion of an activity. Thus gardens provide plentiful opportunities for accessing practical experience as a component of academic learning. Students acquire direct experience when learning gardens are planted directly on the school grounds.

In Chapter 9, "Nurturing Interconnectedness," we described the traditional interplanting technique known as the Three Sisters Guild. This plant guild is a study in relationships: the beans grow upon the corn stalk, while the squash vines provide shade to hold soil moisture, and bacteria living on the bean roots feed the corn. Students often assume that the three plants will compete with one another for resources, and that only one will thrive. Learning about the Three Sisters Guild emphasizes the potentiality for relationships of mutual benefit. Beyond the rich possibilities for academic learning about any of the specific biological interactions present in this plant guild, or the cultural and historical significance of the planting technique, the revelation of mutual benefit is itself a critical insight.

In Chapter 10, "Awakening the Senses," we highlighted meditative "sit spots" as one practice common in many learning gardens. Reading the poetry and journal entries throughout the text demonstrates students' depth of observation and literary skill at expressing emotion. An ability to access the more than intellectual faculties of expression while cultivating academic rigor is fundamental to bringing meaning to existence, for students are more than just minds. We also described the exciting activity of cob construction, a deeply sensorial activity that engages the hands and feet in mixing water, sand, and straw to make a plaster-like substance. While they are having great fun "getting dirty," students are learning about proportions and proper mixing, as well as developing aesthetic sensibility when applying the plaster in free form to the evolving bench. As Principal Tim Lauer notes, graduates frequently return to examine with pride the lasting contribution they made to their old schoolyard habitat. Students build and maintain relationships with place where they have cultivated academic learning and sensorial literacy.

The interdisciplinary and multidimensional possibilities for learning encourage teachers from all subject areas to find their niche in a garden. For example, science teachers can test soil; math teachers can calculate area, rates of change, three-dimensional scale; history teachers can consider changes in land use over time; or art teachers can use the garden as a figure for creative expression.

As illustrated in teacher Deziré Clarke's narrative from practice in Chapter 11, integration across subject areas is possible and particularly effective. Gardens bring learning to life and emphasize the relevance of education. As Superintendent Carlos Garcia points out, "every child eats food"; thus gardens create meaningful connections between everyday experience and learning. Principal Tim Lauer further describes ways in which gardens are integrated throughout the school community and infrastructure.

The unpredictability of the garden shifts focus from teaching toward learning. For example, when learning the parts of a flower from a textbook or diagram, students can see all the parts simultaneously. In the garden, the process gradually unfurls over time fostering a sense of wonder and curiosity as to what may come next. Combining these two learning styles makes for especially effective learning. Similarly learning about cloud forms, or soil horizons, or food webs, all can be made more relevant and tangible in the garden context. Extending the garden by bringing harvest into the cafeteria or connecting with community food banks carries such grounded learning forward into more complex areas of study such as global food systems and social justice issues related to hunger. Connecting with others who are interested in eco-schools and green-schools and who might already be addressing issues of solar power, energy, water, building for sustainability, and so on, helps with linking the work of gardens with other eco-interests in the community.

In the resources section of the Appendix, we provide resources addressing the following: a sample of programs and curriculum models; an inventory of relevant reading material to inspire educators; and a sample of relevant research to demonstrate academic merits of learning gardens.

Theoretical Foundations

Life flows through living soil. In our effort to enliven an educational discourse presently characterized by lifeless mechanistic metaphors, we have introduced living soil as a regenerative alternative framework for thinking about schools as living organisms. Considering living soil as a metaphoric foundation for education underwrites a shift toward sustainability and emphasizes our conviction that truly meaningful learning is enacted in connection with life. Learning gardens planted directly on school grounds provide a practical example of the application of living soil as root metaphor. Like seeds of change that flourish in rich fertile living soil, school gardens and other sustainability education initiatives can be most effective when integrated into an ecological vision of education modeled after living soil.

Learning gardens are effective outdoor extensions of the school that serve as integrative context and contribute to academic achievement even as they create a vibrant space that positively impacts the interrelated problems of diminishment in children's access to natural areas and degradation of soil. Along with providing the physical antidote to nature deficit disorder, gardens positively impact a number of other contemporary crises confronting childhood such as obesity, poor nutrition, and ill health. Supporting the fragrant flowers, beautiful blossoms, and healthful harvest of learning gardens is the living soil, which becomes the basis for an alternative regenerative metaphor to guide sustainability education pedagogy away from mechanistic metaphors.

In the movement toward sustainability we will need both material and metaphoric regeneration: the living soil of learning gardens represents both the physical plane of application and the metaphysical material for developing an ecologically grounded metaphor for education. At a time when education is commonly compared to a "race" and schools are characterized alternatively as "knowledge factories" or "dropout factories," we need a new metaphoric language with which to describe schools as life-serving and life-enhancing entities. Living soil provides a positive basis for developing such a regenerative language even as it serves as home to vibrant biotic life in learning gardens established directly on school grounds. Thus when we cultivate the soil of the school grounds, we also learn a new language with which to describe and understand teaching and learning in an ecological context.

Central to this book has been the contention that schools themselves are living systems and learning gardens are legitimate academic venues and more than just a curricular add-on. We have provided a theoretical foundation to frame learning gardens as a practical application of a vision of schools as life-enhancing ecological communities. A question running throughout the text has been: how can we bring life to schools? We believe that fundamental to the educational enterprise is a responsibility for fostering healthful conditions for all life: life of students, life of nature, life of communities, and life of culture. Thus, rather than preparing students for entry into the alienating and degrading global economy as modern educational imperatives invariably dictate, sustainability education guided via an ecological paradigm supports robust academic learning in tandem with positive contributions to personal, community, and environmental well-being.

As we see grass and asphalt surrounding schools transform into gardens, we emphasize the need to establish secure linkages between the gardens and academic learning. Bringing soil "in sight and in mind" encourages bringing the language of the garden—represented by the metaphor of living soil—to bear within the classroom, rather than bringing the language of the modern school— too often represented by the metaphor of the machine—into the garden.

With this point, we return to a question that has been running through this book: what is education for? To us, the seven principles derived from living soil provide a map toward regenerative education: cultivating a sense of place, fostering curiosity and wonder, discovering rhythm and scale, valuing biocultural diversity, embracing practical experience, nurturing interconnectedness, and awakening the senses. While we see these best articulated in literal contact with living soil in learning gardens on school grounds, these are also elements of good education. When we cultivate a sense of place where we are in relation to broader social, ecological, and economic systems, we are teaching for life. When we foster the flourishing of creativity, curiosity, and wonder, we are enriching children's interest in the world. When we tune attention to discover rhythm and scale all around us, we are deepening the wealth of information available to students. When we connect biological and cultural diversity, we are broadening conversation on difference. When we engage in practical experience, we make learning tangible and stimulate further abstract learning. When we unearth the hidden interconnections that characterize life, we unlock a new understanding of existence. And when we awaken and sharpen our senses, we help students to become aware of the beauty and totality of existence. The seven principles that link pedagogy to pedology combine to offer a regenerative alternative to the dominant modernistic paradigm. The poems, writings, art works, stories and photographs we have shared in this book speak for themselves. At schools across the country, at all age levels, children and youth are grasping the hidden interconnections among food, body, health, community, and soil. Tireless students, educators, parents, community volunteers, university programs, and non-profit organizations are teaming up to bring life to schools in countless ways. They are engaging with real life and soil directly on their school grounds. Along the way they are leading and learning toward sustainability. For this, learning gardens have served as one dynamic venue.

As the preceding pages indicate, learning gardens planted on school grounds are whole-system solutions that contribute positively to numerous indicators of sustainability. In forging multidimensional symbiotic relationships between learning gardens, schools, students, teachers, community members, and living soil, life becomes a central focus of teaching and learning. In learning gardens, living soil and pedagogy surface in dynamic ways to create an ecological landscape of sustainability education.

EPILOGUE

After school has let out for the afternoon, as the sun sinks slowly in the western sky, we are meandering through the winding woodchip paths of learning gardens planted directly on school grounds, admiring the abundant life that surrounds us in each step. Thoughtfully placed bird-baths accent turns along the path, and colorful flowers are interspersed around vegetable beds all of different shapes and sizes. Harvest bins are piled up under the shed roof that serves as an outdoor classroom, water station, and gathering space. The students have gone home for the day and the garden is quiet now, save for the chirps of birds and the rustle of wind. But though the children are no longer present, we can almost hear the squeals of joy that fill this space daily, and we feel the palpable sense of wonder the children repeatedly discover in the endless mysteries of place, plants, soil, and food. The walls of the school that frame the garden appear permeable; less like boundaries that demarcate the end of the outdoors and beginning of the indoors, connection with life flows like water or air back and forth from garden to classroom, animating learning.

As we reflect on learning gardens as a promising form of sustainability education, we know that we are not alone: around the country, hundreds of teachers, parents, students, and community partners are coming together to establish vibrant life-giving spaces directly on school grounds in an effort to place life at the core of education. Also helping in this multifaceted effort are the millions of earthworms and tiny soil micro-organisms that continually replenish the soil that sustains learning gardens. Using a new language for understanding both the process of education and where it is housed, living soil serves as a transformative and regenerative metaphor. *Living* soil is just that—living. In bringing the new frontier of education to the day-to-day ordinary aspects of life, the magnanimity of learning gardens is manifested in ways that only gardens can articulate: an uplifted heart dances with firmly rooted feet; poet and scientist both find their voice; and the human–nature boundaries dissolve. Our hope is that by bringing soil to the foreground in redesigned living landscapes of learning gardens, we may effectively bring life to the center of education.

APPENDIX

Selected Resources and Programs

Sample Programs Connecting Learning Gardens to Academic Subject Areas

The Edible Schoolyards *Program*

The *Edible Schoolyards* (ESY) Program at the Martin Luther King Junior Middle School in Berkeley, California, uses food as a unifying concept and teaches students to grow, harvest, and prepare meals in the school cafeteria. Once an open parking lot, ESY has inspired many programs nationwide encouraging positive interaction with nature.

Website: www.edibleschoolyard.org

The Common Roots *Program*

The *Common Roots* Program in Vermont is an interdisciplinary K–6 grade program that involves community and parents in creation and maintenance of school gardens at school sites. Started by Joseph Kiefer and Martin Kemple (1998) whose primary interest was related to hunger issues, Common Roots has a well-developed spiral curriculum that includes social studies, language arts, mathematics, science, the arts, creative arts, and ecology through gardening and also has community service as an important component of this learning.

Website: www.smartcommunities.ncat.org/success/Common_Roots.shtml

The California School Garden Network

The *California School Garden Network* (CSGN) is a collaborative effort of a number of educational institutions, non-profit organizations, private and government partners committed to enhancing learning through the use of teaching gardens in schools and other community settings. CSGN has developed a comprehensive guide, *Gardens for Learning*, designed to help teachers with new school gardens. The guide is available online at www.csgn.org/page.php?id=36. While specifically written for schools in California, the guide has information and ideas that are useful for schools elsewhere.

Garden-Based Learning *at Cornell University*

The *Garden-Based Learning* Program at Cornell University has developed research-based curriculum materials and activities for schools. The university also provides outreach to schools starting gardens. Information and resources can be found online at http://blogs.cornell.edu/garden/ (Cornell).

LifeLab *Programs at University of California at Santa Cruz*

The *LifeLab* Programs at University of California at Santa Cruz is a premier example of university support for garden-based learning. *LifeLab* supports teachers interested in starting gardens or integrating garden-based learning into their practice. Extensive curriculum resources can be found online at www.lifelab. org/ (UC Santa Cruz).

The Garden Initiative

In Chicago, a program called the *Garden Initiative* has established diverse food and flower gardens at 100 schools throughout the city. For more information about the Chicago gardens see: www.education-world.com/a_curr/curr313.shtml

Sample Readings for Starting and Designing School Gardens

Danks, G. H. (2011). *Asphalt to ecosystems: Design ideas for schoolyard transformation*. Oakland, CA: New Village Press.
This text covers 150 school grounds in over ten countries, with illustrations on how educators at all levels and community members have joined with schoolyard designers to convert asphalt to green spaces. Step-by-step design and school sites are presented.

Gayle, Veronica. (2009). *The learning garden: Ecology, teaching, and transformation*. New York: Peter Lang.
This book narrates the creation of a campus learning garden with student teachers and environmental education students, and uses garden as metaphor to link theory and practice via an ecological model of teaching and learning.

Houghton, E. (2003). *A breath of fresh air: Celebrating nature and school gardens*. Toronto, Canada: Sumach Press.

This book documents with photographs the initiatives detailing school gardens across Toronto. Transformative stories of school gardens are captured showing vividly the practical possibilities.

Sporer-Bucklin, A., & Pringle, R. (2010). *How to grow a school garden: A complete guide for patens and teachers*. Portland, Oregon: Timber Press.

This books offers educators and parents examples of how school gardens can be started in schools. Building on concepts of nutrition, health, and environmental stewardship, the authors expand on their experience in San Francisco to capture the technique and design of school gardens.

Stone, M. K. (2009). *Smart by Nature: Schooling for sustainability*. Healdsburg, CA: Watershed Media.

This book captures the growing sustainability movement in K–12 schools across the United States. A wide range of topics is covered: curriculum, community service-learning, farm to cafeteria innovations, and entire school transformation toward adoption of sustainability initiatives.

Tai, L., Haque, M. T., McLellan, G. K., and Knight, E. J. (2006). *Designing outdoor environments for children: Landscaping schoolyards, gardens, and playgrounds*. New York, NY: McGraw-Hill.

The book presents case studies of outdoor learning environments that are designed for children. Making a case for these spaces to be developmentally appropriate, the authors create a model for teachers to use the environments.

Sample Research: Where is the *Learning* in Learning Gardens?

While there is a fair body of research that is emerging on food habits and nutritional learning as a result of school gardens, below we provide a sample of research related specifically to academic learning (Williams and Dixon, forthcoming).

Aguilar, O. M., Waliczek, T. M., & Zajicek, J. M. (2008). Growing environmental stewards: The overall effect of a school gardening program on environmental attitudes and environmental locus of control of different demographic groups of elementary school children. *HortTechnology*, *18*(2), 243–249.

Alexander, J., North, M. W., & Hendren, D. (1995). Master Gardener classroom garden project: An evaluation of the benefits to children. *Children's Environments*, *12*(2), 123–133.

Beckman, L. L., & Smith, C. (2008). An evaluation of inner-city youth garden program participants' dietary behavior and garden and nutrition knowledge. *Journal of Agricultural Education*, *49*(4), 11–24.

Blair, D. (2009). The child in the garden: An evaluative review of the benefits of school gardening. *Journal of Environmental Education*, *40*(2), 15–38.

Bowker, R., & Tearle, P. (2007). Gardening as a learning environment: A study of children's perceptions and understanding of school gardens as part of an international project. *Learning Environments Research, 10*(2), 83–100.

Brunotts, C. M. (1998). *School gardening: A multifaceted learning tool. An evaluation of the Pittsburgh Civic Garden Center's 'Neighbors and Schools Gardening Together'.* Unpublished master's thesis, Duquesne University.

Castagnino, L. (2005). *Gardens and grade level expectations: The link between environmental education and standardized assessments.* Unpublished master's thesis, Brown University.

Cutter-Mackenzie, A. (2009). Multicultural school gardens: Creating engaging garden spaces in learning about language, culture, and environment. *Canadian Journal of Environmental Education, 14*, 122–135.

Danforth, P. (2005). *An evaluation of the National Wildlife Foundation's Schoolyard Habitat Program in the Houston Independent School District.* Unpublished master's thesis, Texas State University.

Danforth, P. E., Waliczek, T. M., Macey, S. M., & Zajicek, J. M. (2008). The effect of the National Wildlife Federation's Schoolyard Habitat Program on fourth grade students' standardized test scores. *HortTechnology, 18*(3), 356–360.

Dirks, A. E., & Orvis, K. (2005). An evaluation of the Junior Master Gardener Program in third grade classrooms. *HortTechnology, 15*(3), 443–447.

Fleener, A. W. (2008). *The effects of the literature in the garden curriculum on life skills of children.* Unpublished master's thesis, Auburn University.

Harmon, A. H. (1999). *Food system knowledge, attitudes, and experiences.* Unpublished doctoral dissertation, Pennsylvania State University.

Hendren, D. K. (1998). *Evaluation of master gardeners' classroom garden project on youth living in low-income, inner-city neighborhoods of San Antonio.* Unpublished doctoral dissertation, Our Lady of the Lake University.

Karsh, K., Bush, E., Hinson, J., & Blanchard, P. (2009). Integrating horticulture biology and environmental coastal issues into the middle school science curriculum. *HortTechnology, 19*(4), 813–817.

Klemmer, C. D., Waliczek, T. M., & Zajicek, J. M. (2005). Growing minds: The effect of a school gardening program on the science achievement of elementary students. *HortTechnology, 15*(3), 448–452.

Koch, S., Waliczek, T. M., & Zajicek, J. M. (2006). The effect of a summer garden program on the nutritional knowledge, attitudes, and behaviors of children. *HortTechnology, 16*(4), 620–625.

Lekies, K. S., & Sheavly, M. E. (2007). Fostering children's interests in gardening. *Applied Environmental Education and Communication, 6*, 67–75.

Lopez, R., Campbell, R., & Jennings, J. (2008). *Schoolyard improvements and standardized test scores: An ecological analysis.* Boston, MA: Mauricio Gaston Institute for Latino Community Development and Public Policy, University of Massachusetts, Boston.

McArthur, J., Hill, W., Trammel, G., & Morris, C. (2010). Gardening with youth as a means of developing science, work and life skills. *Children, Youth and Environments, 20*(1), 301–317.

Miller, D. L. (2007). The seeds of learning: Young children develop important skills through their gardening activities at a Midwestern early education program. *Applied Environmental Education & Communication, 6*, 49–66.

Morgan, P. J., Warren, J. M., Lubans, D. R., Saunders, K. L., Quick, G. I., & Collins, C. E. (2010). The impact of nutrition education with and without a school garden on knowledge, vegetable intake and preferences and quality of school life among primary-school students. *Public Health Nutrition, 13*(11), 1931–1940.

Morris, J. L. (2000). *The development, implementation, and evaluation of a garden-enhanced nutrition education program for elementary school children.* Unpublished doctoral dissertation, University of California, Davis.

Morris, J. L., & Zidenberg-Cherr, S. (2002). Garden-enhanced nutrition curriculum improves fourth-grade school children's knowledge of nutrition and preferences for some vegetables. *Journal of American Dietetic Association, 102*(1), 91–93.

O'Brien, S. A., & Shoemaker, C. A. (2006). An after-school gardening club to promote fruit and vegetable consumption among fourth grade students: The assessment of social cognitive theory constructs. *HortTechnology, 16*(1), 24–29.

Ozer, E. J. (2007). The effects of school gardens on students and schools: Conceptualization and considerations for maximizing healthy development. *Health Education & Behavior, 34*(6), 846–863.

Pigg, A. E., Waliczek, T. M., & Zajicek, J. M. (2006). Effects of a gardening program on the academic progress of third, fourth, and fifth grade math and science students. *HortTechnology, 16*(2), 262–264.

Robinson, C. W., & Zajicek, J. M. (2005). Growing minds: The effects of a one-year school garden program on six constructs of life skills of elementary school children. *HortTechnology, 15*(3), 453–457.

Sheffield, B. K. (1992). *The affective and cognitive effects of an interdisciplinary garden-based curriculum on underachieving elementary students.* Unpublished doctoral dissertation, University of South Carolina.

Simone, M. F. (2003). *Back to the basics: Student achievement and schoolyard naturalization.* Unpublished master's thesis, Trent University.

Skelly, S. M., & Bradley, J. C. (2000). The importance of school gardens as perceived by Florida elementary school teachers. *HortTechnology, 10*(1), 229–231.

Skelly, S. M., & Bradley, J. C. (2007). The growing phenomenon of school gardens: Measuring their variation and their affect on students' sense of responsibility and attitudes toward science and the environment. *Applied Environmental Education & Communication, 6,* 97–104.

Skelly, S. M., & Zajicek, J. M. (1998). The effect of an interdisciplinary garden program on the environmental attitudes of elementary school students. *HortTechnology, 8*(4), 579–583.

Smith, L. L., & Motsenbocker, C. E. (2005). Impact of hands-on science through school gardening in Louisiana public elementary schools. *HortTechnology, 15*(3), 439–443.

Sobaski, C. (2006). *An investigation of interactivity at the Michigan 4-H Children's Garden.* Unpublished master's thesis, University of Delaware.

Thorp, L. G. (2001). *The pull of the earth: An ethnographic study of an elementary school garden.* Unpublished doctoral dissertation, Texas A&M University.

Waliczek, T. M. (1997). *The effect of school gardens on self-esteem, interpersonal relationships, attitude toward school, and environmental attitude in populations of children.* Unpublished doctoral dissertation, Texas A&M University.

Waliczek, T. M., Bradley, J. C., & Zajicek, J. M. (2001). The effect of school gardens on children's interpersonal relationships and attitudes toward school. *HortTechnology, 11*(3), 466–468.

Waliczek, T. M., Logan, P., & Zajicek, J. M. (2003). Exploring the impact of outdoor environmental activities on children using a qualitative text data analysis system. *HortTechnology, 13*(4), 684–688.

Waliczet, T. M., & Zajicek, J. M. (1999). School gardening: Improving environmental attitudes of children through hands-on learning. *Journal of Environmental Horticulture, 17*(4), 180–184.

BIBLIOGRAPHY

Abbott, S. (1983). *Womenfolks: growing up down south.* New York, NY: Houghton Mifflin.

Ableman, M. (1998). *On good land: The autobiography of an urban farm.* San Fransisco: Chronicle Books.

Ableman, M. (2005). Raising whole children is like raising good food: Beyond factory farming and factory schooling. In Z. Barlow & M. K. Stone (Eds.), *Ecological literacy: Educating our children for a sustainable world* (pp. 175–183). San Fransisco: Sierra Club.

Abram, D. (1996). *The spell of the sensuous.* New York: Vintage Books.

Agyeman, J. (2005). *Sustainable communities and the challenge of environmental justice.* New York, NY: NYU Press.

Agyeman, J., & Evans, B. (2004). 'Just sustainability': the emerging discourse of environmental justice in Britain. *The Geographical Journal, 70*(2),155–164.

Alexander, J., North, M. W., & Hendren, D. (1995). Master Gardener classroom garden project: An evaluation of the benefits to children. *Children's Environments, 12*(2), 123–133.

Anderson, J. (2009). *Tongue-tied no more: Learning Gardens and social justice.* Unpublished master's research project, Portland State University, Oregon.

Arendt, H. (1971). *The life of the mind: thinking.* New York, NY: Harcourt.

Arnold, T. (2004). *Catalina Magdalena Hoopensteiner Wallendiner Hogan Logan Bogan was her name.* New York, NY: Scholastic.

Barnhardt, R. (2007). Creating a place for indigenous knowledge in education: the Alaska Native Knowledge Network. In Smith, G. & Gruenewald, D. (Eds.), *Place-based education in the global age: Local diversity* (pp. 113–153). New York, NY: Lawrence Erlbaum Associates.

Barrett, T. J. (1947) *Harnessing the earthworm.* Boston: Bruce Humphries.

Barrs, R., Lees, E. E., & Philippe, D. (2002). *School ground greening: A policy and planning guidebook.* Canada: Evergreen.

Basso, K. H. (1996). *Wisdom sits in places: Landscape and language among Western Apache.* Albuquerque: University of New Mexico Press.

Berlyne, D. E. (1960). *Conflict, arousal, and curiosity.* New York: McGraw-Hill.

Berry, W. (1970a). *Farming: A handbook.* New York, NY: Harcourt Brace Jaconovich.

Berry, W. (1970b). *A continuous harmony: Essays cultural and agricultural.* Washington, DC: Shoemaker and Howard.

Berry, W. (1990). *What are people for?* San Francisco: North Point Press.

Berry, W. (1991). *The unforseen wilderness: Kentucky's red river gorge.* Emeryville, CA: Shoemaker & Hoard.

Berry, W. (1992). Speech delivered November 11, 1994 at the 23rd annual meeting of the Northern Plains Resource Council, Billings, MT. http://sustainabletraditions.com/2010/10/wendell-berry-17-rules-for-a-sustainable-local-community/ (retrieved August 5, 2011).

Berry, W. (2004). *Citizenship papers.* Washington, DC: Shoemaker & Howard.

Bhatt, C. P. (April 1990). The Chipko Andolan: Forest conservation based on people's power. *Environment and Urbanization, 2*(1), 7–18.

Blair, D. (2009). The child in the garden: An evaluative review of the benefits of school gardening. *Journal of Environmental Education, 40*(2), 15–38.

Bowers, C. A. (1997). *The culture of denial: Why the environmental movement needs a strategy for reforming universities and public schools.* Albany: State University of New York Press.

Bowers, C. A. (2000). *Let them eat data: How computers affect education, cultural diversity, and the prospects of ecological sustainability.* Athens: University of Georgia Press.

Brink, L., & Yost, B. (2004). Transforming inner-city school grounds: lessons from learning landscapes. *Children, Youth and Environments, 14*(1), 208–232.

Bronfenbrenner, U. (1979). *The ecology of human development: experiments by nature and design.* Cambridge, MA: Harvard University Press.

Brown, P. (2009). *Curious garden.* New York, NY: Little Brown Books.

Bucklin-Sporer, A., & Pringle, R. (2010). How to grow a school garden: A complete guide for parents and teachers. Portland, OR: Timber Press.

Cajete, G. A. (2001). Indigenous education and ecology: Perspectives of an American Indian educator. In J.A. Grim (Ed.), *Indigenous traditions and ecology: The interbeing of cosmology and community* (pp. 619–638). Cambridge, MA: Harvard University Press.

Cajete, G. A. (2005). American Indian epistemologies. *New Directions for Student Services, 109,* 69–78. http://onlinelibrary.wiley.com.proxy.lib.pdx.edu/doi/10.1002/ss.155/pdf (retrieved December 16, 2010).

Capra, F. (1975). *The Tao of Physics: An Exploration of the Parallels Between Modern Physics and Eastern Mysticism.* Berkeley, CA: Shambhala.

Capra, F. (1996). *The web of life: A new scientific understanding of life.* New York, NY: Anchor.

Capra, F. (2002). *The hidden connections: Integrating the biological, cognitive, and social dimensions of life into a science of sustainability.* New York, NY: Doubleday.

Carson, R. (1956). *The sense of wonder.* New York, NY: Harper.

Carson, R. (1962). *Silent spring.* New York, NY: Houghton Mifflin.

Castagnino, L. (2005). *Gardens and grade level expectations: The link between environmental education and standardized assessments.* Unpublished master's thesis, Brown University.

Chawla, L. (2006). Learning to love the natural world enough to protect it. *Barn. Norsk Senter for Barnesorskinning, 2,* 57–68.

Chiro, G. D. (1996). Nature as community: The convergence of environment and social justice. In W. Cronon (Ed.), *Uncommon ground: Toward reinventing nature* (pp. 298–320). New York, NY: Norton.

Clarke, D. C. (2010). *Ethnobotany: A year of schoolyard learning. Curriculum Review.* Unpublished master's research project, Portland State University, Oregon.

Cobb, E. (1969). The ecology of imagination in childhood. In P. Shepard & D. McKinley (Eds.), *The subversive science: essays toward an ecology of man.* Boston: Houghton Mifflin.

Cocks, M. (2006). Biocultural diversity: Moving beyond the realm of 'indigenous' and 'local' people. *Human Ecology, 34*(2), 185–197.

Code, L. (2006). *Ecological thinking: The politics of epistemic location.* Oxford: Oxford University Press.

Commeyras, M. (1995). What can we learn from students' questions? *Theory into Practice,* *34*(2), 101–106.

Cuban, L. (1990). Reforming again, again, and again. *Educational Researcher, 19*(1), 3–13.

Cutter-Mackenzie, A. (2009). Multicultural school gardens: creating engaging garden spaces in learning about language, culture, and environment. *Canadian Journal of Environmental Education, 14,* 122–135.

Damrow, C. B. (2005). *"Every child in a garden": radishes, avocado pits, and the education of American children in the twentieth century.* Unpublished Doctoral Dissertation, University of Wisconsin-Madison.

Danforth, P. (2005). *An evaluation of the National Wildlife Foundation's Schoolyard Habitat Program in the Houston Independent School District.* Unpublished master's thesis, Texas State University.

Dardis, G., Parajuli, P., & Williams, D. (2008). *Curriculum development and preparation in and for the learning gardens.* Report submitted to Oregon Community Foundation, Portland, Oregon.

DeLind, L. B. (2006). Of bodies, place, and culture: Re-situating local food. *Journal of Agricultural and Environmental Ethics, 19,* 121–146.

Desmond, D., Grieshop, J., & Subramaniam, A. (2002). *Revisiting garden-based education in basic education: philosophical roots, historical foundations, best practices and products, impacts, outcomes and future directions.* Food and Agriculture Organization, United Nations International Institute for Educational Planning.

Dewey, J. (1910). *How we think.* Boston: Heath.

Dewey, J. (1916). *Democracy and Education: An introduction into the philosophy of education.* New York: MacMillan.

Dewey, J. (1925). *Experience and nature.* La Salle, IL: Open Court.

Dewey, J. (1938). *Experience and education.* New York, NY: Macmillan.

Dillard, A. (1974). *Pilgrim at Tinker Creek.* New York, NY: HarperCollins.

Dillard, A. (1999). *For the time being.* New York, NY: First Vintage Books.

Dirks, A. E., & Orviz, K. (2005). An evaluation of the junior master gardener program in third grade classrooms. *HortTechnology, 15*(3), 443–447.

Doris, E. (1991). *Doing what scientists do: Children learn to investigate their world.* Portsmouth, NH: Heinemann.

Dyment, J. (2005). Green school grounds as sites for outdoor learning: barriers and opportunities. *International Research in Geographical and Environmental Education, 14*(1), 28–45.

Esteva, G., & Prakash, M. S. (1998). *Grassroots postmodernism.* New York, NY: St Martin's Press.

Fagot, A., Burrows, N., & Williamson, D. (1999). The public health epidemiology of type 2 diabetes in children and adolescents: a case study of American Indian adolescents in the southwestern United States. *Clinica Chimica Acta, 286,* 81–95.

Fisher, J. (2000) *Pass the celery, Ellery!* New York, NY: Stewart, Tabori & Chang.

Fleischman, P. (1999). *Seed folks.* New York, NY: Harper Collins.

Freire, P. (1970). *Pedagogy of the oppressed.* New York, NY: Herder and Herder.

Friedman, T. L. (2005). *The world is flat: A brief history of the twenty-first century.* New York, NY: Farrar, Straus and Giroux.

Fukuoka, M. (1978). *The one straw revolution.* New York, NY: New York Review Books.

Fusco, D. (2001). Creating relevant science through urban planning and gardening. *Journal of Research in Science Technology, 38*(8), 860–877.

Gadamer, H. (2001). *Gadamer in conversation,* trans. Richard Palmer (from Gadamer, 1993), New Haven, CT: Yale University Press.

Gandhi, M.K. (1953). *Towards new education.* Ahmedabad, India: Navjivan Press.

Gardner, H. (1999). *Intelligence reframed: Multiple intelligences for the 21st century.* New York, NY: Basic Books.

Gaylie, V. (2009). The Learning Garden: Ecology, teaching, and transformation. New York, NY: Peter Lang.

Gliessman, S. R. (1984). An agroecological approach to sustainable agriculture. In W. Jackson, W. Berry, & B. Colman (Eds.), *Meeting the expectations of the land: Essays in sustainable agriculture and stewardship* (pp. 160–171). San Fransisco: North Point Press.

Graham, H., & Zidenberg-Cherr, S. (2005). California teachers perceive school gardens as an effective nutritional tool to promote healthful eating habits. *Journal of the American diabetic association. 105*(11), 1796–1800.

Green, T. F. (1971). *The activities of teaching.* Cambridge, MA: Harvard University Press.

Greene, M., & Griffiths, M. (2003). Feminism, philosophy, and education: Imagining public spaces. In N. Blake, P. Smeyers, R. Smith, & P. Standish (Eds.), *The Blackwell guide to the philosophy of education* (pp. 73–92). Oxford: Blackwell.

Gruenewald, D. A. (2003). The best of both worlds: A critical pedagogy of place. *Educational Researcher, 32*(4), 3–12.

Gruenewald, D. A. (2008). Place-based education: Grounding culturally responsive teaching in geographical diversity. In D. A. Gruenewald & G. A. Smith (Eds.), *Place-based education in the global age: Local diversity* (pp. 137–154). New York, NY: Taylor & Francis Group.

Harmon, D. (2002). *In light of our differences: How diversity in nature and culture makes us human.* Washington, DC: Smithson Institution Press.

Harper, J. (2004). *Four boys named Jordan.* New York, NY: G.P. Putnam.

Hawken, P. (2007). *Blessed unrest: How the largest movement in the world came into being and why no one saw it coming.* New York, NY: Penguin.

Hawkins, J. (2010). Hands-On Pastoral Education using Clergy Sustaining Agriculture. www.hopecsa.org/documents/2010HOPECSAsyllabus.pdf

Hayden-Smith, R. (2006). *Soldiers of the soil: a historical review of the United States school garden army.* Davis, CA: University of California 4-H Center for Youth Development.

Hedley, A., Ogden, C., Johnson, C., Carroll, M., Curtin, L., & Flegal, K. (2004). Overweight and obesity among U.S. children, adolescents, and adults. *Journal of the American Medical Association, 291*(23), 2847–50.

Hemenway, T. (2000). *Gaia's garden: A guide to home-scale permaculture.* White River Juntion, VT: Chelsea Green.

Henkes, K. (1991). *Chrysanthemum.* New York, NY: Harcourt.

Holmgren, D. (2002). *Permaculture: Principles and pathways beyond sustainability.* Hepburn, Victoria, Australia: Holmgren Design Services.

Hyams, E. (1976). *Soil and civilization.* London: Harper & Row.

Jeavons, J. (1974). *How to grow more vegetables: And fruits, nuts, berries, grains, and other crops than you ever thought possible on less land than you can imagine.* Berkeley: Ten Speed Press.

Jones, V. (2007). *Where are low-income and minority greens in the media?* http://gristmill.grist.org/story/2007/5/19/154942/683

Jones, V. (2008). *The green-collar economy: How one solution can fix our two biggest problems.* New York, NY: Harper Collins.

Kawagley, A. O., & Barnhardt, R. (1999). Education indigenous to a place: Western science meets native reality. In G. A. Smith & D. R. Williams (Eds.), *Ecological education: On weaving education, culture, and the environment* (pp. 117–142). Albany: State University of New York Press.

Kellert, S. R. (1996). *The value of life: biological diversity and human society.* Washington, DC: Island Press.

Kellert. S. R. (2005). *Building for life: Designing and understanding the human-nature connection.* Washington, DC: Island Press.

Kiefer, J., & Kemple, M. (1998). *Digging deeper: Integrating youth gardens into schools & communities*. Vermont: Food Works.

King, M. L. (1967) Commencement address at Oberlin College June 1965. www.oberlin.edu/external/EOG/BlackHistoryMonth/MLK/CommAddress.html (retrieved August 5, 2011).

Klemmer, C. D., Waliczek, T. M., & Zajicek, J. M. (2005). Growing minds: the effect of a school gardening program on the science achievement of elementary students. *HortTechnology, 15*(3), 448–452.

Klindienst, P. (2006). *The earth knows my name: Food, culture, and sustainability in the gardens of ethnic Americans*. Boston: Beacon Press.

Knowles, R. L. (1981). *Sun rhythm form*. Cambridge, MA: MIT Press.

Knowles, R. L. (1999). The solar envelope. www-bcf.usc.edu/~rknowles/sol_env/sol_cnv.html (retrieved December 2, 2010).

Koch, S., Waliczek, T. M., & Zajicek, J. M. (2006). The effect of a summer garden program on the nutritional knowledge, attitudes, and behaviors of children. *Hort Technology, 16*(4), 620–625.

Kohlstedt, S. G. (2008). "A better crop of boys and girls": the school gardening movement, 1890–1920. *History of Education Quarterly, 48*(1), 58–93.

Korten, D. (2006). *The great turning: from empire to earth community*. Bloomfield, CT: Kumarian Press

Krapfel, P. (1999). Deepening children's participation through local ecological investigations. In G. A. Smith and D. R. Williams (Eds.), *Ecological education in action: On weaving education, culture, and the environment* (pp. 57–64). New York, NY: State University of New York Press.

Kumar, S. (2002). *You are therefore I am*. Devon, England: Green Books.

Kunstler, J. H. (1993). *The geography of nowhere: The rise and decline of America's man-made landscape*. New York, NY: Simon and Schuster.

Landa, E., & Feller, C. (Eds.). (2010). *Soil and culture*. New York, NY: Springer.

Latham, G. (1996). Fostering and preserving wonderment. *Australian Journal of Early Childhood 21*(1), 12–15.

Leopold, A. (1949). *Sand county almanac*. London: Oxford University Press.

Logan, W. (1996). *Dirt: The ecstatic skin of the earth*. London: Norton & Company.

Lopez, B. H. (1978). *Of wolves and men*. New York, NY: Scribner.

Lopez, R., Campbell, R., & Jennings, J. (2008). *Schoolyard improvements and standardized test scores: an ecological analysis*. Research Briefs. Mauricio Gaston Institute for Latino Community Development and Public Policy, University of Massachusetts, Boston.

Lopez, R., Campbell, R., & Jennings, J. (2008). The Boston Schoolyard Initiative: A public-private partnership for rebuilding urban play spaces. *Journal of Health Politics, Policy and Law, 33*(3), 617–638.

Louv, R. (2008). *Last child in the woods: Saving our children from nature-deficit disorder*. Chapel Hill, NC: Algonquin Books.

Lovelock, J. (2000). *Gaia: a new look at life on earth*. Oxford: Oxford University Press.

Maathai, W. M. (2006). *Unbowed: A memoir*. New York, NY: Anchor Books.

Macy, J. (2009). The great turning. www.joannamacy.net/thegreatturning.html (retrieved on December 25, 2010).

Maffi, L. (2001). *On biocultural diversity: linking language, knowledge, and the environment*. Washington, DC: Smithsonian.

Maffi, L. (2005). Lingusitic, cultural, and biological diversity. *Annual Review of Anthropology, 34*, 599–617.

Maffi, L. (2007). Biocultural diversity and sustainability. In Pretty, J. N., Ball, A., Benton, T., Guivant, J., Lee, D. R., Orr, D., Pfeffer, M. and Ward, H. (Eds.), *Sage Handbook on Environment and Society* (pp. 267–278). Thousand Oaks, CA: Sage.

Malone, K., & Traner, P. (2003). School grounds as sites for learning: making the most of environmental opportunities. *Environmental Education Research, 9*(3), 283–303.

Martinez, D. (1996). First people, firsthand knowledge. *Sierra* (Nov/Dec), 50–51, 70–71.

Masumoto, D. M. (2003). *Four seasons in five senses: Things worth savoring.* New York, NY: Norton.

McLaren, P., & Houston, D. (2004). Revolutionary ecologies: Ecosocialism and critical pedagogy. *Journal of the American Educational Studies Association, 36*(1), 27–45.

McNeill, J. R., and Winiwarter, V. (Eds.). (2006). *Soil and societies: Perspectives from environmental history.* Isle of Harris: The White Horse Press.

Mollison, B. (1988). *Permaculture: A designer's manual.* Tyalgum, Australia: Tagari Press.

Mollison, B. (1990). *Permaculture: A practical guide for a sustainable future.* Washington, DC: Island Press.

Montgomery, D. R. (2007). *Dirt: The erosion of civilizations.* Berkeley: University of California Press.

Morgan, P. J., Warren, J. M., Lubans, D. R., Saunders, K. L., Quick, G. I., & Collins, C. E. (2010). The impact of nutrition education with and without a school garden on knowledge, vegetable intake and preferences and quality of school life among primary-school students. *Public Health Nutrition, 13*(11), 1931–1940.

Morris, J. L., Neustadter, A., & Zidenberg-Cherr, S. (2001). First-grade gardeners more likely to taste vegetables. *California Agriculture, 55*(1), 43–46.

Muir, J. (1911). *My first summer in the Sierra.* New York, NY: Dover Books.

Nabhan, G. P. (1997). *Cultures of habitat: on nature, culture, and story.* Washington, DC: Counterpoint.

Nabhan, G. P. (2002). *Coming home to eat: The pleasures and politics of local foods.* New York, NY: Norton.

Nabhan, G. P. (2004). *Why some like it hot: food, genes, and cultural diversity.* Washington, DC: Island Press.

Noddings, N. (1992). *The challenge to care in schools: an alternative approach to education.* New York, NY: Teachers College Press.

Opdal, P. (2001). Curiosity, wonder, and education seen as perspective development. *Studies in Philosophy of Education, 20,* 331–344.

Orr, D. W. (1992). *Ecological literacy: Education and the transition to a postmodern world.* Albany: State University of New York Press.

Orr, D. W. (1994). *Earth in mind: On education, environment, and the human prospect.* Washington, DC: Island Press.

Ozer, E. (2007). The effects of school gardens on students and schools: conceptualization and considerations for maximizing healthy development. *Health Education and Behavior, 34,* 846–863.

Parajuli, P. (2001). Learning from ecological ethnicities: Toward a plural political ecology of knowledge. In J. A. Grim (Ed.), *Indigenous traditions and ecology: The interbeing of cosmology and community* (pp. 559–590). Cambridge, MA: Harvard University Press.

Parajuli, P. (2002). Sustainability Partnership Model. Leadership in Ecology, Culture, and Learning Program, Portland State University.

Parajuli, P. (2006). Learning Suitable to Life and Livability: Innovation through Learning Gardens. *Connections: The Journal of the Coalition of Livable Future* 8(1), Spring, 6–7.

Parajuli, P., & Williams, D. (2005). Learning Gardens Laboratory: Health, multiculturalism, and academic achievement. A report submitted to the Portland City Council, Portland, Oregon.

Pollan, M. (2006). *The omnivore's dilemma: A natural history of four meals.* New York, NY: Penguin Books.

Posey, D. A. (Ed.) (1999). *Cultural and spiritual values of biodiversity.* London and Nairobi: Intermediate Technology Publication and UNEP.

Postman, N. (1994). *The disappearance of childhood*. New York, NY: Knopf.

Prakash, M. S. (1994). Think locally, act locally. *Holistic Education Review* 7(4), 50–56.

Pretty, J. (2008). How do biodiversity and culture intersect? Plenary paper for Conference "Sustaining Cultural and Biological Diversity In a Rapidly Changing World : Lessons for Global Policy". Organized by American Museum of Natural History's Center for Biodiversity and Conservation, IUCN-The World Conservation Union/Theme on Culture and Conservation, and Terralingua.

Pretty, J., Adams, B., Berkes, F., de Athayde, S., Dudley, N., Hunn, E., Maffi, L., Milton, K., Rapport, D., Robbins, P., Sterling, E., Stolton, S., Tsing, A., Vintinnerk, E., & Pilgrim, S. (2009) The Intersections of Biological Diversity and Cultural Diversity: Towards Integration. Conservat Soc (serial online) 7, 100–112. www.conservation-andsociety.org/text.asp?2009/7/2/100/58642 (retrieved August 6, 2011).

Ratcliffe, M. M. (2007). *The effects of school gardens on children's knowledge, attitudes, and behaviors related to vegetable consumption and ecoliteracy*. Doctoral dissertation, Tufts University, Friedman School of Nutrition Science and Policy.

Redford, K. H., & Brosius, J. P. (2006). Diversity and homogenization in the endgame. Editorial. *Global Environmental Exchange 16*, 317–319.

Rilke, R. M. (1934). *Letters to a young poet*. New York, NY: Norton.

Robinson, C. W., & Zajisek, J. M. (2005). Growing minds: the effect of a one year school garden program on six constructs of life skills of elementary school children. *HortTechnology, 15*(3), 453–457.

Rosenberg, M. B. (2003). *Nonviolent communication: A language of life*. Encinitas, CA: PuddleDancer Press.

Sachs, C. E. (1996). *Gendered fields: Rural women, agriculture, and environment*. Boulder, CO: Westview Press.

Sauvé, L., Berryman, T., & Brunelle, R. (2007). Three decades of international guidelines for environment-related education: A critical hermeneutic of the United Nations discourse. *Canadian Journal of Environmental Education, 12*, 33–45.

Schmitt, F. F., & Lahroodi, R. (2008). Epistemic value of curiosity. *Educational Theory 58* (2), 125–148.

Sewell, L. (1995). The skill of ecological perception. In M. Gomes, A. Kanner, & T. Roszak. (Eds.), *Ecopsychology: Restoring the earth and healing the mind*, (pp. 201–215). Berkeley: The University of California Press.

Shiva, V. (1993). *Monocultures of the mind*. London: ZeD Books.

Shiva, V. (2000). *Stolen harvest: The hijacking of the global food supply*. Cambridge, MA: South End Press.

Shiva, V. (2005). *The polarised world of globalization*. www.zcommunications.org/the-polarised-world-of-globalisation-by-vandana2-shiva (retrieved August, 2, 2011).

Shiva, V. (2008). *Soil not oil: Environmental justice in a time of climate crisis*. Cambridge, MA: South End Press.

Sipos, Y., Battisti, B., & Grimm, K. (2008). Achieving transformative sustainability learning: Engaging head, hands and heart, *International Journal of Sustainability, 9*(1), 68–86.

Skelley, S. M., & Bradley, J. C. (2007). The growing phenomenon of school gardens: measuring their variation and their affect on students' sense of responsibility and attitudes toward science and the environment. *Applied Environmental Education and Communication, 6*(1), 97–104.

Smith, G., & Gruenewald, D. (Eds.). (2008). *Place-based education in the global age: Local diversity*. New York, NY: Lawrence Erlbaum Associates.

Smith, G., & Sobel, D. (2010). *Place-and community-based education in schools*. New York, NY: Routledge.

Smith, G.A. (2002). Place based education: learning to be where we are. *Phi Delta Kappan, 83*(8), 584–594.

Smith, G. A., & Williams, D. R. (Eds.). (1999). *Ecological education in action: On weaving education, culture, and the environment*. Albany: State University of New York Press.

Snyder, G. (1990). *The practice of the wild*. San Francisco: North Point.

Sobel. D. (2004). *Place-based education: Connecting classrooms and communities*. Great Barrington, MA: The Orion Society.

Sobel, D. (2008). *Childhood and nature: Design principles for educators*. Portland, ME: Stenhouse.

Sterling, S. (2001). *Sustainable education: Re-visioning learning and change* (Schumacher Briefings No. 6). Devon, England: Green Books.

Stone, M. K. (2009). *Smart by nature: Schooling for sustainability*. Healdsburg, CA: Watershed Media.

Stone, M. K., & Barlow, Z. (2005). *Ecological literacy: educating our children for a sustainable world*. San Francisco: Sierra Club Books.

Subramaniam, A. (2002). *Garden-based learning in basic education: a historical review*. Davis, CA: University of California, 4-H Center for Youth Development.

Swimme, B., & Berry, T. (1992). *The universe story: An autobiography from planet Earth*. San Francisco: Harper and Row.

Taylor, S. (2009). Seasonal calendar of events, Sunnyside Environmental School, Portland Public Schools, Portland, OR.

Taylor, A. F., Kuo, F. E., & Sullivan, W. C. (2001). Coping with ADD: the surprising connection to green play settings. *Environment and Behavior, 33*(1), 54–77.

Thorp, Laurie. (2006). *Pull of the earth: Participatory ethnography in the school garden*. Lanham, MD: AltaMira Press.

Tolle, E. (1999). *The power of now: a guide to spiritual enlightenment*. Novato, CA: New World Library.

Trelstad, B. (1997). Little machines in their gardens: a history of school gardens in America, 1891–1920. *Landscape Journal, 16*(2), 161–173.

Tschannen-Moran, M., & Uline, C. (2008). The walls speak: the interplay of quality facilities, school climate, and student achievement. *Journal of educational administration, 46*(1), 55–73.

UNESCO. (2010). *Education for sustainable development*. www.unesco.org/en/esd/

Varadarajan, S. (2005, August 2). *Book review: But the world's still round*. www.hindu.com/br/2005/08/02/stories/ 2005080200381500.htm

van Oudenhoven, F. J. W., Mijatovic, D., & Eyzaguirre, P. B. (2010). Bridging the managed landscapes: The role of traditional (agri) culture in maintaining diversity and resilience of social-ecological systems. In Belair, C. *et al.* (Eds.), *Secretariat of the Convention of Biological Diversity in Socio-Ecological Production of Landscapes* (pp. 8–21). UNU-IAS.

Waters, A. (2009, Feburary 19). No lunch left behind. www.nytimes.com/2009/02/20/opinion/20waters.html?_r=1&scp=1&sq=no+lunch+left+behind&st=nyt

Wells, N. M. (2000). At home with nature: effects of greenness on children's cognitive functioning. *Environment and Behavior, 32,* 775–795.

Wessel, T. (1999). *Reading the forested landscape*. Woodstock, VT: Countryman Press.

Williams, D. R. (1993). The unity of learning and living for ecological sustainability: Gandhi's educational philosophy in practice. *Holistic Education Review 6*(3), 18–22.

Williams, D. R. (1995). Think ecologically, act locally: On becoming native to a place. *Holistic Education Review 8*(2), 52–54.

Williams, D. R. (2008). Listening to nature: Cultivating ecological literacy through Learning Gardens. *Oregon English Journal, 30*(1), 12–15.

Williams, D. R. (2008). Sustainability Education's gift: learning patterns and relationships. *International Journal of Education for Sustainable Development, 2*(1), 45–50.

Williams, D. R., & Brown, J. D. (2011). Living soil and sustainability education: linking pedagogy with pedology. *Journal of Sustainability Education*.

Williams, D. R., & Dixon, S. P. (forthcoming). *Garden-based learning: Synthesis of research.* A report to the Spencer Foundation.

Wilson, E. O. (1984). *Biophilia: The human bond with other species.* Cambridge, MA: Harvard University Press.

Winne. M. (2008). *Closing the food gap: Resetting the table in the land of plenty.* Boston: Beacon Press.

Young, J. (2001). *Exploring natural mystery.* Duvall, WA: Owlink Media.

INDEX

Note: Page numbers in *italics* are for tables, those in **bold** are for figures.

Abbott, Shirley 61
Abernathy Elementary, Portland Public School District 28–9
Ableman, Michael 175, 176
Abram, David 148, 152
abstract ideas, privileging of 7, 8
academic achievement 5, 24, 147, 199; competitive approach to 4–5, 8; gap 5, 39
academic learning xiv, 22, 23, 162, 194, 196–9
academic outcomes 23, 24; indirect 36–7
Access Academy K–8, Portland Public School District 161, 162–78
African Green Belt movement 135
age of ecology 90
age of Enlightenment 90
Agyeman, Julian 65, 135
Alexander, Jacquelyn 24
Alice Fong Yu K–8, San Francisco Unified School District 27–8, 180
Anderson, J. 144
apples 116–17
Aptos Middle School, San Francisco Unified School School District 65
Arendt, Hannah 76
Artemisia ludoviciani 113
assimilation, cultural 109

attitudinal changes 23, 24
autonomy 7–8, 63

Barnhardt, Ray 60
beans: and Three Sisters Garden 36, 70, 141–4, 198; *see also* germinating beans experiment
bees 106, 111
Berry, Wendell xvii, 14, 38, 57, 61, 62, 73, 174
Bibbeau, Matt 182–3
Bindweed 112–13
biocultural diversity xiv, 14, 19, 47, 48, 52–3, 79, 94, 105–19, 197–8, 200; ecological perspective 106–9; examples from learning gardens 110–18; in soils of living gardens 105–6
biodynamic gardening 93
biointensive gardening 43, 93
biological diversity xii, 45, 48, 105, 106, 107–9, 110, 111, 118, 119, 197, 201; *see also* biocultural diversity
biomass 4, 42, 43, 50, 53, 99
biomimicry 197
bioregions 13, 15, 24, 33, 47, 195, 196
birds in flight, chevron pattern of 93
Blair, Dorothy 24
bodily experiences 32, 147, 148; *see also* senses

body–mind relationship 145, 147, 148
books 81, 82; picture 165
Boston public high schools 59, 65
Bowers, C.A. 14, 44
Bronfenbrenner, U. 193
Brosius, J. Peter 109–10
Brown, Jonathan 29
Brown, Peter, *Curious Garden* 68–9
Bucklin-Sporer, Arlene 27
Bureau of Education 23
butterflies 97

Cajete, Gregory A. 136, 144
California School Garden Network
    204–5
camouflage and wildlife 128–9
Capra, Fritjof 14, 35, 42 133, 138, 145,
    157, 193
carbon markets 135
care: curriculum of 139; ethic of 20, 35,
    60, 73, 124, 197
Carson, Rachel 47, 75, 76, 84, 137
Cartesian paradigm 6
Castagnino, Laura 24
celebration of life xiv
character webs 168
Chawla, Louise 67
chemical use in agriculture 137
The Chicago Garden Initiative 25, 205
Chicago High School of Agricultural
    Sciences, Chicago Public School
    District 25
Chicago Public School District 25, 59
Children and Nature Network 9
Chipko movement 135
chronological time 91, 95, 103
chronos 91
citizenship, soil-based 66
Clarke, Deziré 161, 163–78, 192–3,
    197, 199
class differences: in academic
    achievement 5, 39; in food access 37
classroom constitution 166
climate 58, 173
climate change 10–11, 35, 62
cob construction 154–6, 182, 198
Cocks, Michelle 107, 109
comfrey 111
Common Roots project, Vermont
    67, 204
community 66, 67, 73, 134, 164, 169
community collaboration 26–7, 29–30,
    59–60, 61–2, 174, 191, 193

community gardens 25, 26, 28, 80,
    117–18
community history 59
community level impact 24
community mapping 171
companion planting 36, 142; *see also*
    Three Sisters Garden
competitiveness 4–5, 8, 45, 63, 95
compost 43–4, 49, 99; carbon and
    nitrogen ratios 100; greens and
    browns 100; as intergenerational
    gift 50, 51; lesson plan **51**
compost making 49–53
composting 33, 37, 198; food scrap 26,
    37, 155, 185
composting in place 43
connecting gardens with lunchroom
    185–6
conserving seed 100–2, 114–15
context, importance of 65, 66
continuity of educational experience 139
continuity of place 139
Convolvulus Arvensis 112–13
corn 36, 71–2; and Three Sisters Garden
    141–4, 198
Cornell University 205
creativity 7, 19, 109, 165, 201
critical pedagogy of place 64–5
critical thinking 35, 41, 80, 81, 175, 197
Cuban, Larry 5
culinary and agrarian arts xiv
culinary traditions 30, 39
cultural awareness 21
cultural capital 7
cultural dimension of food 30, 36,
    39, 106
cultural diversity xiv, 22, 39, 45, 48, 94,
    105, 106, 108–9, 110, 118–19, 197,
    201; *see also* biocultural diversity
cultural groups, partnerships between
    and among 19, 20, 29
cultural knowledge 115
cultural memory 61, 114, 118
cultural traditions xiv, 26, 36, 94, 107
culture 58, 106, 107; and ecology xiv,
    38–9, *see also* biocultural diversity;
    local 7; soil and 57
culture–nature: divide xii, 22, 125, 134;
    relationship xii, 12, 15, 20, 21
curiosity and wonder xiv, 197, 200, 201;
    distinguished 78; ecological perspective
    78–81; examples from learning gardens
    81–8; fostering 14, 36, 41, 46, 47, 52,

75–88; loss of 6, 8; sensory stimulation and 147
Curious Garden (Brown) 68–9
curriculum 3, 6; of care 139; continuity of 139; homogenization of 7, 8; integrated 30, 31–3, 165, 175, 186–7, 191, 192–3, 199; for learning gardens 30–2; nationalization of 23; subject area standards 18, 23, 30; *see also* seasonal curriculum

Danforth, Phillip 24
Darwin, Charles 7
de-contextualization of learning 6, 7, 8, 47
de-paving events 79–80
decomposers 37, 50, 86, 100
decomposition xii, 50, 52, 84–6, 100
DeLind, Laura B. 115
Denver Public School District 26–7, 59
Denver Public Schools Nutrition Services 27
Denver Urban Gardens (DUG) 26, 59, 117–18
Descartes, René 7, 90; *see also* Cartesian paradigm
designing the landscape 28, 179, 183–4
designing school gardens 19, 38, 42, 44, 176–7, 183–4
Desmond, Daniel 22
Dewey, John 11, 14, 22, 67, 122, 124–6
Diehl, Don 27
Dillard, Annie 123, 156
direct experience xiv, 81–2, 198
Dirks, Amy 24
discrimination 121, 122, 124, 126, 132
diversity: valuing of 33, 192; *see also* biocultural diversity; biological diversity; cultural diversity
Dixon, Phillip 23
dwellers 60, 73

Eagleton Elementary, Denver Public School District 27
Earth, as single living organism 91
earth sciences *31*, 32
earthly seasonal rhythms 90–1
ecological milieu 21–2
ecological partnerships 19, 20, 21
ecological/environmental injustice xiii, 10–11
ecology 32; and culture xiv, 38–9; *see also* biocultural diversity

ecology–economy divide 22
economic relationships 19, 20
Ecotrust, Portland 28
edge effect 33
edge experience 156, 157
Edible Schoolyard, Berkeley 154, 180, 204
educational reforms xii, xiii, 4–5, 63, 95
efficiency 5, 94, 96; using natural rhythms and scales 99
Empty Bowls concept 65
engagement with life 122, 123, 126, 131, 132
Enlightenment ideas 7–8, 90
environmental ethics 19
environmental justice 134–5; *see also* ecological/environmental injustice
environmental movement 62–3, 137
equity 64, 65
Esteva, Gustavo 14, 61
ethic of care 20, 35, 60, 73, 124, 197
ethics 19, 20
ethnobotany 166, 173, 192
etymological relationships 13
evolution 7–8, 138
experience xiv, 7, 14, 39, 47, 48, 53, 65, 67, 120–32, 147, 200; active 124; continuity of 122, 124, 126; Dewey's concept of 124–6; direct xiv, 81–2, 198; examples from learning gardens 128–31; Gandhi's philosophy and role of 126–7; passive/vicarious 124, 127, 131; quality of 125; restoration of 384
Eyzaguirre, Pablo B. 107

Facebook 11
facts and figures, beyond memorization of 16
fair trade 20
Fairview Elementary, Denver Public School District 26–7, 59, 117, 118
family trees 166
Farmers-in-Residence programs 28, 59, 190–1
fast-food 149
Ferguson, Sara 163
fertilizers 137
field guides 173
field journals 26
Fleischman, Paul, *Seedfolks* 18, 19, 168–9
food 32, 105–6; appreciation of 24; contamination of 62; costs 35, 37; cultural dimension of 30, 36,

39, 106; fast- 149; historic dimension
of 36; local 30, 35, 57–8; novel 16,
24; sources of 9, 35
food access 36–7
food banks 35, 37, 129, 199
Food-based Ecological Education Design
(FEED) program 29, 179–80, 181
food connections activity *112*
food consciousness 9
food deserts 36–7
food gardens 9, 24, 30, 39, 191, 194;
see also vegetable gardens
food scrap composting 26, 37, 155, 185
food security/insecurity xiii, 9, 37,
65, 139
forests, cutting 135
free trade 20
Freire, Paulo 66, 67
Friedman, Thomas 10, 62
Friends of the Urban Forest 188
frost 121, 122

Gadamer, Hans-Georg 122
Gaia hypothesis 91, 136
Gandhi xvii, 82, 123, 124, 126–7;
Nai Talim or *New Education* 127
Garcia, Carlos 162, 189–92, 192, 193–4,
196, 197, 199
garden art 27, 163
garden based learning 18, 22; history
22–3; outcomes 23–4; rationales 23;
research 23–4
Garden-based Learning, Cornell University
205
garden coordinators 181
The Garden Initiative (Chicago) 25, 205
garden journals 171, 173–4
Garden of Wonders 28–9
gardens: as ecological milieu 21–2; as
outdoor learning centers 193
Gardens for Learning 205
Gardner, Howard 139
generations *see entries under*
intergenerational
genetic modification 137
geography 18, 23
geometry 102, 131, 197
germinating beans experiment 86–8
Global Environmental Change 109–10
global thinking 47, 63
globalization 9, 10, 22, 57, 58, 59, 60, 61,
62–3, 64, 65, 110
Green Revolution 137

Green, Thomas F. 75, 83
Greene, Maxine 5
Grieshop, James 22
Griffiths, Morwenna 5
groundedness in place 58–61, 73
Growing Gardens, Portland 28, 29
Gruenwald, David 64

Hacienda, Portland 28
Hahn, Thich Nhat xiii
haiku 116–17
Hamlin, Julia 183
hands, using 1, 90, 123, 126, 127
Harmon, David 105, 118–19
Harvest of the Month 30, 143, 191
harvest rituals 72
Hawkins, Jeffery 104
head, hand, and heart (3Hs) 30,
123–4, 127
health 32, 40, 117; of the planet 117;
soil 43, 53, 137–8; student xii, 24, 117;
see also ill-health
health science 18, 23, *31*
healthy eating xii, 30, 33, 154
healthy food choices 9
hearing 8, 151
Hendren, Deborah 24
herb spiral 176–7, 197
herb tea 172–3
high-stakes testing 6
high-status knowledge 7, 81, 82, 127
historic dimensions of food 36
historical background 22–3
history 61; local/community 59
holistic education 165, 187
holistic relationships 8, 21
Holmgren, David 33
homelessness 35, 37, 65
homogenization 48, 106, 107, 109–10,
131; cultural 109; of curriculum
and learning 7, 8
Houston Independent School
District 24
humus 13, 43, 59, 83
hunger 35, 37, 59, 65, 117, 199
hunger banquets 116
hydroponics 44

identity 73
ill-health xiii, 9, 37, 200
imagination 36, 41
immigrant communities 29, 39, 110, 117;
see also migrant workers

indigenous communities 36, 60–1, 106, 107, 137; interaction with trees 135; rituals and ceremonies 94, 107, 113; Three Sisters planting system 36, 141; traditional education 136
indirect learning outcomes 36–7
indirect nature experiences 81, 82
individual level impact 24
individualism 7–8, 63, 64, 196
infrastructure support 179, 180–1, 189, 190–1
integrated curriculum 30, 31–3, 165, 175, 186–7, 191, 192–3, 199
intelligence(s), multiple 139
inter-cultural partnership 19, 20, 29
inter-economic partnership 19, 20
inter-species partnership 19, 20
interconnectedness xii, 14, 47, 48, 53, 64, 133–45, 193, 198, 200, 201; ecological perspective 136–8; examples from learning gardens 140–4; learning gardens as nexus of 139; as relationships 133–5; senses and understanding of 147, 157; social dimension of 134, 135
interdisciplinary nature: of garden education 18, 19, 20, 198; of permaculture principles 33; see also integrated curriculum
intergenerational gifts to soil 37, 50, 51, 53
intergenerational learning 30, 67, 127
intergenerational relationships 26, 29, 97, 115, 149; see also intra- and inter-generational partnership
internet 81, 138
intra- and inter-generational partnership 19, 20, 29

Jackson, Wes 61
Jaramillo, Abby 28
Jeavons, John 93

kairos 91
kairotic time 91, 97
Kawagley, Oscar 60
Kellert, Stephen R. 81
Kelly Elementary, Portland Public School District 29
Kemple, Martin 204
keyhole beds 38
Kiefer, Joseph 204
King, Martin Luther 135

kinship with soil xii
Klemmer, Cynthia D. 24
Klindiest, Patricia 110
knowledge: cultural 115; de-contextualized 6, 7, 65; high-status 7, 81, 82, 127; low-status 7, 123
knowledge factories 200
Knowles, Ralph L. 89, 91
Kohlstedt, Sally Gregory 22, 23
Kolouch, Gigia 27
Krapfel, Paul 36, 114
Kumar, Satish 14, 137–8

LaCourse, Isabel 163
Lahroodi, Reza 77
land ethic 61
land-use practices 61; indigenous 36
landscape and mindscape 8, 118, 123, 128
Lane Middle School, Portland Public School District 29, 99, 102, 128, 143–4
language 109
language arts 18, 23, 31
Lauer, Tim 162, 178, 179–87, 192, 193, 196, 197, 198, 199
Leadership in Ecology, Culture, and Learning (LECL) program xvii
learning: academic xiv, 22, 23, 162, 194, 196–9; and curiosity and wonder 76, 77, 79; de-contextualization of 6, 7, 8, 47; by doing 123, 124, 127; experiential see experience; homogenization of 7, 8; intergenerational 30, 67, 127; and life see life and learning; and linear time 95–6; mechanistic model of 6, 8, 45, 95; monocultural 39; multicultural 21, 30, 39, 67, 110; multidisciplinary 30; multiple levels of 16; multisensory 30; service 30, 35, 127
Learning Gardens Laboratory (LGL), Portland 29, 31, 39, 128
Learning Landscapes, Denver 27, 117
learning outcomes 23, 24; indirect 36–7
learning symbiosis 140–1
legacy garden 176–8
legitimate academic venues 15, 200
Leki, Pete 25, 26
Leopold, Aldo 61, 136–7
LeVan, Angela 181, 182
Lewis Elementary, Portland Public School District 29, 153, 154, 162, 178–87
Lewis Outdoor Education Center, Portland 155, 178, 184

life, engagement with 122, 123, 126, 131, 132
life cycles 32, 93–4, 157
life and learning: connection 11–13, 15, 23; divide 22
life sciences *31*, 32, 191
LifeLab program, University of California at Santa Cruz 180, 190, 205
life's marvels and mysteries xiv
light 94; lunar 93, 94
linear time 95–6
linguistic diversity 108, 118
literature 18, 23
living soil: and interconnectedess 134; as literal medium xii–xiii, 42–3, 45, 58, 96, 118, 131, 133, 139, 200; and local place 57–8; as metaphor xiii, 13, 14, 42–9, 54, 58, 79, 88, 90, 103, 138, 139, 145, 199–200, 202; nitrogen cycle and 140; rhythm and scale as attributes of 89–90, 99; and sensory stimulation 149
local food 30, 35, 57–8
local–global dichotomy 47, 63
local–global interconnections 47, 58–9, 61, 62, 72
local histories 59
local place 57–8, 60, 62–4, 66
local thinking 47, 63–4
Lopez, Barry 38
Louv, Richard 9
Lovelock, James 91
low-income communities 27, 36–7, 39, 65
low-status knowledge 7, 123
lunar cycle 93–4
lunar light 93, 94
Luther Burbank Middle School, San Francisco Unified School District 28

Maathai, Wangaari 120, 135
Madison High School, Portland Public School District 28
Maffi, Luisa 108
manual labor 7, 123, 127
mapping the community 171
mapping the gardens 32–3
Marchise, Terrie 190
Martin Luther King Jr. Middle School, San Francisco Unified School District 28, 154, 180, 204
Martinez, Dennis 112
Master Gardeners 28

Masumoto, David 149, 150–1
mathematics 18, 20, 23, 24, *31*, 92, 100, 131, 192; *see also* geometry
meaning(s): making 121, 126, 147; networks of 39; restoration of 38
mechanical and industrial scale 6–7, 8, 94, 95
mechanistic metaphors 5, 6, 8, 11, 42, 44, 45, 54, 94, 95, 103, 200
mechanistic perspectives 136, 137
medicinal value of plants 30, 111, 113, 176
membership with soil 12
memory 147, 150; cultural 61, 114, 118
metaphor(s): of living soil xiii, 13, 14, 42–9, 54, 58, 79, 88, 90, 103, 138, 139, 145, 199–200, 202; mechanistic 5, 6, 8, 11, 42, 44, 45, 54, 94, 95, 103, 200
Mickelberry-Morris, Madelyn 163, 171
migrant workers 189–90, 194
Mijatovic, Dunja 107
mind–body relationship 145, 147, 148
mindscape: and landscape 8, 118, 123, 128; redesigning the 44
Minto tribe, Alaska 60
model of school gardens 24
Mollison, Bill 21, 33, 141
monocultural learning 39
monoculture, global 62
monoculture of the mind 33, 45, 106, 109
monocultures of plants 45, 109, 137
Monsanto 137
moon 32, 47, 91, 93–4, 99
moral development 24
more-than-human worlds 18, 20, 21, 38, 40
Muir, John 16
multicultural gardens 117–18, 198
multicultural learning 21, 30, 39, 67, 110
multidisciplinary learning 30
multiple levels of learning 16
multisensory learning 30
mycrorrhizal networks 140, 144
mythologies 81

Nabhan, Gary Paul 38, 115
Nai Talim or *New Education* (Gandhi) 127
names/naming 167
National Wildlife Foundation 25
Native Americans *see* indigenous communities

native gardens 30; *see also* Sabin
Community Native Garden
native plants 163, 173, 192
Native Wormwood 113
natural rhythms 14, 47, 48, 52, 90–102,
103–4, 147
nature: direct experiences of xiv, 81–2;
indirect experiences of 81, 82; listening
to 16, 153; symbolic experiences of 81;
as teacher 174
nature–culture: divide xii, 22, 134, 135;
relationship xii, 12, 15, 20, 21
nature deficit disorder 9, 12, 139, 200
nature in the neighborhood 37
nested systems xiv, 157
nitrogen cycle 140
nitrogen fixation 134, 140–1, 145
No Child Left Behind (NCLB) xii, 5, 6, 8,
10, 63, 81, 139
No Child Left Inside Initiative 10
no-till polycultural farming 43
Noddings, Nel 138, 139
North, Mary-Wales 24
Novak, Andrew 27
nutrition 18, 23, 24, 30, 32; inadequate
37, 200

Obama, Michelle, First Lady xii
obesity xiii, 9, 117, 139, 200
observation 79, 82, 83, 84, 146–7,
195, 196
Opdal, Paul 76, 77
oral traditions/histories 7, 59, 81, 97
Oregon Green Premier School 185
Orr, David 14, 60, 65–6, 138
Orvis, Kathryn 24
outdoor learning centers, gardens as 193
outdoor learning experiences xii
Oxfam America Hunger Banquet 116
Ozer, Emily 24

Parajuli, Pramod xvii, 19–20, 29,
39, 179
parochialism 63, 64
partnership model of sustainability 19–21
partnerships 179–80, 193, 196
patterns xiv, 91–3, 97, 99, 197
peaches 149, 150
pedagogical principles 14, 46–9, 51–3;
for sustainability education xiii
pedagogy 3; in the community 19;
critical 64–5; learning gardens xiii, 19,
42, 44, 104; sustainability 19–21

pedology, principles linking pedagogy
and xiv, 46–9
permaculture 30, 33, 38, 141, 176
permaculture principles 30, 33, 195
personal development 32
pesticides 137
physical activity xii
physical rhythms 90
physical sciences *31*, 32
picture books 165
place 164; continuity of 139; critical
pedagogy of 64–5; ecological
perspective 61–4; examples from
learning gardens 67–73; groundedness
in 58–61, 73; and identity 73; local
57–8, 60, 62–4, 66; sense of xiv, 14,
32–3, 46, 47, 51, 57–74, 147, 174,
192, 196, 201; as unrecognised in
education 65–6
place-based education xiv, 64, 66–74,
164, 175
plant functions 111–13
plant growth patterns 93–4
plant guilds 36, 141; *see also* Three Sisters
Garden
play 36
poetic and critical texts 18
polycultural planting 43, 137, 141; *see also*
Three Sisters Garden
Portland Public School District 28–30, 59,
79–80, 161, 178
Portland State University (PSU) 29, 30,
179–80
Postman, Neil 88
pottery bowls 65
practical experience *see* experience
practice and theory, intersection of xiv,
xv, 21
Prakash, Madhu 14, 63
Pretty, Jules 108, 109
principal's perspective 161, 162, 179–86
Pringle, Rachel 27
progress 96, 123
pumpkins *see* squash/pumpkins

questions 77, 78, 81, 83, 88

Race to the Top xii, 5, 8, 63, 81, 95
racial differences: in academic achievement
5, 39; in food access 37
rain gardens 30
rainwater harvesting 37, 188
raised beds 38

reading 20
reciprocal relationships 39, 45, 57, 58, 62, 73, 136, 150
recycling 33, 38
Redford, Kent H. 109–10
reflection 121, 122, 125, 126, 132, 198
refugee communities 26, 27, 29, 39, 117, 118
relationships: appreciation of 79; see also interconnectedness; partnership(s); reciprocal relationships
religious activity 94
research 23–4
residents 60
respect 45, 46
reviving food knowledge 9
Rhode Island 24
rhythm(s) 92, 93, 197, 200, 201; artificial 89; examples from learning gardens 96–102; and living soil 90; natural 14, 47, 48, 52, 90–102, 103–4, 147; and patterned systems 93; physical 90
Rilke, Rainer Maria 14, 88
rural–urban relationships 20

Sabin Community Native Garden, Portland 162, 163, 174
Sabin Edible Garden, Portland Public School District 162
Sabin Parent Teacher Association 163
sacred spots see sit spots
San Francisco 37, 59, 65, 116
San Francisco Commission on the Environment 188–9, 191
San Francisco Green Schoolyards Alliance (SFGSA) 27, 188
San Francisco Unified School District (SFUSD) 27–8, 65, 161, 162, 180, 188–92; Bond Greening Program 188
scale 14, 47–8, 52, 89–90, 103, 147, 197, 200, 201; examples from learning gardens 97, 98–9, 102; and living soil 90, 99; mechanical and industrial 6–7, 8, 94, 95; in modern Western society 48, 89; optimal 92; orders of 91–2, 93
Schmitt, Frederick F. 77
school–community divide 22
school level impact 24
Schools Uniting Neighborhood (SUN) program 162–3
science 18, 20, 23, 24, 28, 31, 32, 191, 192
science talks 167–8

seasonal calendar 97–8, 197
seasonal curriculum 99–102; integration with subject standards 30, 31–3
seasonal cycles and rhythms 32, 90–1, 96
seasonal variation 94, 157
Seattle, Chief 136
secret garden spots 36
Seed Folks (Fleishman) 18, 19, 168–9
Seed Kids 19, 169–70, 174
seeding memory 114
seeds: heritage or heirloom 102, 114–15; saving 100–2, 114–15
self-sufficiency 12, 63
senses: awakening the 14, 47, 48–9, 53, 146–57, 198, 200, 201; ecological perspective 149–51; examples from learning gardens 152–6; stimulation of 8, 147
sensory awareness 82, 147, 151
sensory capacity 49, 146, 147, 148, 151–2
sensory engagement 83, 146
service-learning 30, 35, 127
Sewell, Laura 146, 152
SFUSD Bond Greening Program 188
Shiva, Vandana 10, 14, 61, 137
sight 8, 150, 151
sit spots 82, 152, 153–4, 198
Slow Food Denver 26, 27, 117
Slow Food, Portland 28
smell(s) 8, 149–50, 151, 156, 157
Smith, Gregory 64, 67
Snyder, Gary 61, 74
Sobel, David 66, 67
social development 24, 32
social groups, partnership between and among 19, 20
social justice/injustice 10, 35, 64, 65, 135, 137, 199
social media 11, 81
social safety net 10, 139
social studies 31
soil 33, 82–3, 84–6, 120–1; abuse and disguise of 42, 44; analysis 59; biocultural diversity in 105–6; building 33, 38, 43, 49–53; chemical use in 137; and culture 57; diversity 106, 108, 118; edge experiences 156; erosion 10, 43; health 43, 53, 137–8; intergenerational gifts to 37, 50, 51, 53; kinship with xii; and life 13–14; membership with 12; negative connotation of 42; personalities 43; science 83; tilth 120; see also living soil; pedology

soil-based citizenship 66
soil kids xii
soil and land stewardship xii
songs 81, 97, 115, 127
soul 137
sound 150, 151
soup kitchens xiii, 131
space xiv, 90, 92–3, 96, 98, 99
specialization 7, 48
species: extinction 111; partnership
   among 19, 20
speed 94, 95
spiderwebs 97
spiral garden design 38, 176–7, 197
spirituality 36
square foot gardening 102, 197
squash/pumpkins 36, 37, 70, 92, 113,
   114; and Three Sisters Garden 70,
   141–4, 198
stacking functions technique 33
standardization 5, 80, 109
standards, subject area 18, 23
status *see* high-status knowledge;
   low-status knowledge
Steele Elementary, Denver Public School
   District 27
Steelville, Missouri 65
stone soup 115–16
stories 81, 97, 115, 127
storyline projects 68–73
straw bale mulch gardening 38
student health xii, 24, 117
subject area: integration 30, 31–3, 165,
   175, 186–7, 191, 192–3, 199; standards
   18, 23
Subramaniam, Aarti 22
sun 47, 91, 93, 99
Sunnyside Environmental School K–8,
   Portland Public School District 29,
   68–73, 84–8, 116, 142–3, 153
superintendent's perspective 161, 162,
   189–92
sustainability xii, 78, 200; education
   as 53; partnership model of 19–21;
   as an umbrella term 3–4
sustainability education xiii, 3, 8, 19, 21,
   79, 103, 104, 199, 200; discourse
   of 4; diversity and 108; engagement as
   key element of 131; living
   soil metaphor and 42, 53, 88;
   place-based learning and 73; whole
   systems thinking and 176
sustainability pedagogy 19–21

symbiotic relationships 134, 138, 140–1,
   144, 145
symbolic experiences of nature 81

Tarkington K–8, Chicago Public School
   District 25
taste 8, 147, 151, 156, 157
Taylor, Sarah xvii
teacher's perspective 161, 163–78, 192–3
technical drawings 171, 197
technology 8, 81, 127, 138, 151
terrariums 85
testing 5, 80; high-stakes 6; multiple
   choice 5, 6, 80
theory and practice, intersection of xiv,
   xv, 21
think ecologically, act locally xiv, 47,
   62–4
think globally 47, 63
think locally 47, 63
thinking big 6–7, 96
Thompson, Valerie 163
Three Sisters Garden 36, 70, 141–4,
   145, 198
time xiv, 90, 92, 93, 98, 197;
   chronological 91, 95, 103; cyclical
   sense of 52; as fluid and recursive
   104; kairotic/event-based 91, 97;
   linear 95–6
touch 8, 150, 151, 156
transnational corporations 137
transportation 35, 37
trees 99, 135, 196–7

uncertainty 78, 80
United Nations 3
United States School Garden Army 23
University of Alaska 60
University of California at Santa Cruz
   180, 190, 205
University of Colorado, Denver 27
Urban Sprouts, San Francisco 27, 28, 190

van Oudenhoven, Frederik J.W. 107
van Zandt, Steven 86
vegetable gardens xii, 25, 129, 131
Vernon Elementary, Portland Public
   School District 29
vertical gardening 38
Vestal Elementary, Portland Public School
   District 79–80
Village Building Convergence, Portland
   154, 182, 183

Wagner, Rebecca xvi
Waliczek, Tina M. 24
wartime school gardens 22, 23
waste audits 37, 154
water cycle 43
Waters, Alice 154, 180
Waters K–8, Chicago Public School
    District 25–6
Waterworks, Portland 28
weather patterns 94
web of relationships 35
Wellness in the Garden 27
Wessels, Tom 61
White House xii
White Sagebrush 113
whole-systems 8, 11, 33, 61, 139,
    176, 201

whole-systems design solutions
    19, 42
wildlife: camouflage 128–9; guilds 37;
    habitat 37, 111
Williams, Dilafruz 23, 29, 47, 62, 64
Wilson, Edward O. 152
wonder *see* curiosity and wonder
the world is flat 10, 62
worm bins 38
writing 20, 24, 192

Young, Jon 82

Zajicek, Jayne M. 24
Zoroastrians 94